Lecture Notes in Mathematics

A collection of informal reports and seminars
Edited by A. Dold, Heidelberg and B. Eckmann, Zürich

Series: Mathematisches Institut der Universität Bonn
Adviser: F. Hirzebruch

290

Don Bernard Zagier
Mathematisches Institut der Universität Bonn,
Bonn/Deutschland

Equivariant Pontrjagin Classes and Applications to Orbit Spaces

Applications of the G-signature Theorem to Transformation Groups, Symmetric Products and Number Theory

Springer-Verlag
Berlin · Heidelberg · New York 1972

AMS Subject Classifications (1970): 57-02, 57B99, 57D20, 57E15, 57E25, 58G10

ISBN 3-540-06013-8 Springer-Verlag Berlin · Heidelberg · New York
ISBN 0-387-06013-8 Springer-Verlag New York · Heidelberg · Berlin

Offsetdruck: Julius Beltz, Hemsbach/Bergstr.

INTRODUCTION

This volume contains an assortment of results based on the Atiyah-Singer index theorem and its corollaries (the Hirzebruch signature and Riemann-Roch theorems and the G-signature theorem). Because the applications of this theory have so wide a scope, the reader will find himself involved with characteristic classes, finite group actions, symmetric products of manifolds, and number theory of the naive sort. On top of this, he may feel that the level of presentation is swinging up and down in a dizzying fashion. I hope I may prevent, or at least relieve, his seasickness by a few preliminary remarks about the level and content of the material.

The results ought to be comprehensible to a working topologist (or even a good graduate student) who is not necessarily a specialist on the Atiyah-Singer theorem. The non-expert should thus not be put off by references in the introduction to esoteric theorems of Thom, Atiyah-Singer, and the like, nor be further discouraged when he finds that even the first section of Chapter One throws no more light on these matters. Background material is, in fact, included, but it has been postponed to the second section so that the main theorems of the chapter can be collected together at the beginning for reference. A similar course has been pursued in Chapter Two.

Aside from this point, I should perhaps mention that a much more thorough treatment of the required background on characteristic classes, index theorems and group actions can be found in the notes [21] (if they ever appear), which also contain a further selection of results in the same direction as those of this volume, and to some extent complement it (overlap of results has been minimized).

We now give a summary of the contents of the volume.

Hirzebruch defined for a differentiable manifold X a characteristic class

$$L(X) \in H^*(X;\mathbb{Q}) \qquad (1)$$

which, on the one hand, is determined by the Pontrjagin class of X, and, on the other, determines the signature of X. Thom showed how to define L(X) when X is only a rational homology manifold.

Our goal in Chapter I will be to generalize this to a definition of an "equivariant L-class"

$$L(g,X) \in H^*(X;\mathbb{C}) \qquad (g \in G) \qquad (2)$$

for a rational homology manifold X with an orientation-preserving action of a finite group G. Apart from their intrinsic interest, these classes will make it possible to compute the L-class (in Thom's sense) of certain rational homology manifolds.

In the differentiable case, we define L(g,X) by

$$L(g,X) = j_! L'(g,X), \qquad (3)$$

where $L'(g,X) \in H^*(X^g;\mathbb{C})$ is the cohomology class appearing in the G-signature theorem, $j: X^g \subset X$ is the inclusion of the fixed-point set, and $j_!$ is the Gysin homomorphism. We then show (§3) that

$$\pi^* L(X/G) = \sum_{g \in G} L(g,X), \qquad (4)$$

where $\pi: X \to X/G$ is the projection onto the quotient. Since the map $\pi^*: H^*(X/G;\mathbb{C}) \to H^*(X;\mathbb{C})$ is injective, this completely calculates the L-class in Thom's sense for the simplest sort of rational homology manifold, namely the quotient of a manifold by a finite group.

We then imitate Milnor's reformulation of Thom's definition to give a definition of L(g,X) for rational homology manifolds which agrees with (3) for differentiable manifolds. Formula (4) still holds, and indeed can be extended to calculate the new equivariant L-classes for orbit spaces.

As an example (§6), we evaluate $L(g,P_n\mathbb{C})$ for g acting linearly on $P_n\mathbb{C}$, and use this to calculate the L-class of $P_n\mathbb{C}/G$ for G a finite, linear action (the result had already been obtained by Bott). Also, by studying the behaviour of the formula for $L(g,P_n\mathbb{C})$, we can formulate various conjectures about the nature of the classes L(g,X).

The whole of Chapter II, which occupies half of the volume, is an application of the result (4). We take X to be the nth Cartesian product of a manifold M, and G the symmetric group on n letters, acting by permutation of the factors. The quotient $X/G = M(n)$, called the nth symmetric product of M, is a rational homology manifold if $\dim M = 2s$, and we can apply (4) to calculate its L-class. The complete result is complicated, but displays a simple dependence on n, namely

$$L(M(n)) \quad = \quad j^*(Q^{n+1}G), \tag{5}$$

where j is the inclusion of $M(n)$ in $M(\infty)$ and $Q, G \in H^*(M(\infty))$ are independent of n. Moreover, the "exponential" factor Q is very simple and, so to speak, independent of M: we have

$$Q \quad = \quad Q_s(\eta), \tag{6}$$

where $\eta \in H^*(M(\infty))$ is a class defined canonically by the orientation class $z \in H^*(M)$, and where

$$Q_s(\eta) \quad = \quad 1 + (3^{-s})\eta^2 + (5^{-s} - 2\cdot 9^{-s})\eta^4 + \ldots \tag{7}$$

is a power series depending only on s. The "constant" factor G, though known, is very much more complicated,* and is only of any real use for manifolds with very simple homology. In §13, we compute it in two cases: for $M = S^{2s}$, where we find

$$G \quad = \quad \frac{Q_s(\eta) - \eta Q_s'(\eta)}{Q_s(\eta)^2 - \eta^2}, \tag{8}$$

and for s=1, i.e. M a Riemann surface. In the latter case, M(n) is a smooth (indeed, complex) manifold and $Q_s(\eta) = \eta/\tanh \eta$ is the Hirzebruch power series. In this case $L(M(n))$ was known (the Chern class of M(n) was found by Macdonald), so we can check our main theorem.

We can restate (5) without the class G, in the form

$$j^*L(M(n+1)) \quad = \quad Q_s(j^*\eta)\cdot L(M(n)) \tag{9}$$

(here the first j denotes the inclusion of M(n) in M(n+1)). This is reminiscent of the relationship between the L-classes of a manifold A and submanifold B (namely $j^*L(A) = L(\nu)\cdot L(B)$, where $j: B \subset A$ and ν is the normal bundle). A direct interpretation is impossible because the inclusion $M(n) \subset M(n+1)$ does not have a "good" normal bundle, even in Thom's extended sense (this follows from (9) and the fact that the power series $Q_s(t)$ does not split formally as a finite

---

* Since this volume was written, I have found a simpler expression for G involving the (finitely many) multiplicative generators of $H^*(M(\infty);\mathbb{Q})$ rather than the additive basis described in §7. However, this will appear--if at all--elsewhere.

product $\Pi_{j=1}^{s}(x_j/\tanh x_j)$ for $s>1$). However, equation (9) seems to suggest strongly the existence of some more general type of bundle (possibly analogous to the "homology cobordism bundles" defined by Maunder and Martin for the category of $\mathbb{Z}$-homology manifolds) which would be appropriate to inclusions of rational homology manifolds and which would possess L-classes. There is some reason to believe that "line bundles" of this type would be classified by maps into the infinite symmetric product $S^{2s}(\infty)$. Since this space (by the theorem of Dold and Thom) is a $K(\mathbb{Z},2s)$, such "line bundles" over X would be classified by a "first Chern class" in $[X,K(\mathbb{Z},2s)] \cong H^{2s}(X;\mathbb{Z})$.

We end Chapter II by calculating $L(g,M(n))$, where g is an automorphism of a 2s-dimensional manifold M of finite order p (then g acts on $M(n)$ via the diagonal action on M^n). We find that (9) is replaced by

$$j^*L(g,M(n+p)) \quad = \quad \{p^{-s}\,\eta^{p-1}\,Q_s(p^s\eta)\}\,L(g,M(n)) \tag{10}$$

if p is odd, and has no analogue at all if p is even. Again we have the possibility of checking our results in the two-dimensional case, this time by taking $M=S^2$ and comparing with the results of Chapter I on the Bott action on $P_n\mathbb{C} = S^2(n)$.

In Chapter III we make explicit calculations with the G-signature theorem on certain simple manifolds ($P_n\mathbb{C}$ with the Bott action, Brieskorn varieties, and related manifolds), and relate them to the number-theoretic properties of finite trigonometric sums such as

$$\operatorname{def}(p;q_1,\dots,q_{2n}) \quad = \quad (-1)^n \sum_{j=1}^{p-1} \cot\frac{\pi j q_1}{p}\cdots\cot\frac{\pi j q_{2n}}{p} \tag{11}$$

(where $p \geq 1$, $q_1,\dots,q_{2n}$ integers prime to p). We prove that (11) is a rational number whose denominator divides the denominator of the Hirzebruch L-polynomial L_n (i.e. 3 for n=1, 45 for n=2, etc.). We also prove a new "reciprocity law" for the expressions (11), both by elementary methods and--in two different ways--by specializing the G-signature theorem.

Although it is not made apparent here, there is a close tie between the results of Chapter III and the result in Chapter I on the L-class of $P_n\mathbb{C}/G$ (cf. [21]).

* * *

The research described in this volume took place in Oxford and Bonn during the years 1970-71; I would like to thank both of these institutions, as well as the National Science Foundation and the Sonderforschungsbereich Theoretische Mathematik der Universität Bonn for financial support. Above all, my thanks go to Professor Hirzebruch, who taught me the little I know and much more.

* * *

Notation is fairly standard, except that for want of italics we have underlined symbols occurring in the text (not, however, Greek or capital letters or expressions containing more than one letter: thus we would write "let $\underline{a}$ be a point of a set A" but "then λ equals $e^{2\pi ix}$."). We use $|A|$ to denote the number of elements of a finite set A.

References to the bibliography have been made in the normal way, by the use of appropriate numbers in square brackets; an exception is the reference Spanier [38] which like every one else we refer to simply as "Spanier."

The numbering of theorems, propositions, lemmata and equations starts afresh in each section. The symbol §3(10) denotes equation (10) of section 3.

TABLE OF CONTENTS

CHAPTER I: L-CLASSES OF RATIONAL HOMOLOGY MANIFOLDS

In his famous paper "Les classes caractéristiques de Pontrjagin des variétés triangulées" ([39]), R. Thom showed that it is possible to define a Hirzebruch L-class $L(X) \in H^*(X;\mathbb{Q})$ (or equivalently a rational Pontrjagin class) for a rational homology manifold X, in such a way as to obtain the usual L-class if X possesses the structure of a differentiable manifold. This definition rested on the possibility of making precise the notion of a rational homology submanifold of X with a normal bundle in X, and showing that X has enough such submanifolds to represent all of its rational homology. The definition was later simplified by Milnor [31], who observed that it is easy to give a definition of a "submanifold with trivial normal bundle" agreeing with the usual concept if X is differentiable (such a manifold is $f^{-1}(p)$, where f is a map from X to a sphere and p is a point of the sphere in general position), and that it follows from the work of Serre [37] that there are also enough of these more special submanifolds to represent all of $H_*(X;\mathbb{Q})$ (indeed there are just enough, i.e. a one-one correspondence; in Thom's definition each homology class was represented by many submanifolds and one had to check consistency as well as sufficiency). Nevertheless, the definition remained essentially an existence proof rather than a procedure for actually computing $L(X)$, and as a result the definition has remained of relatively little intrinsic interest and has been most important for its use in proving facts about the ordinary L-class or rational Pontrjagin class (e.g. that this is the same for two differentiable manifolds with the same underlying PL structure).

There is, however, one especially simple type of rational homology manifold, namely a quotient space X/G of a smooth manifold X by an orientation-preserving action of a finite group G, and for such a space it is possible to give a formula for the L-class in terms of the action of G on X by using the G-signature theorem of Atiyah and Singer. This formula will be given in §1 and proved in §3. An illustration of it will be given in §6, where we calculate $L(X/G)$ for $X = P_n\mathbb{C}$ and G a product of finite cyclic groups acting linearly on X; the L-class of this space had already been calculated by Bott using a different method. A much more difficult application is to the L-class of the nth symmetric product M(n) of a manifold M (here $X = M^n$ and G is the symmetric group on n letters, acting on X by permutation of the factors); this will be carried out in Chapter II.

In the formula for L(X/G), certain cohomology classes $L(g,X) \in H^*(X;\mathbb{C})$ occur, defined for each $g \in G$ and such that $L(id,X) = L(X)$. Their definition in the differentiable case is based on the G-signature theorem and thus requires a knowledge of certain normal bundles and of the action of g on these bundles, so that it depends very heavily on the differentiable structure. However, it is possible to define these "equivariant L-classes" also when X is only a rational homology G-manifold in a manner exactly parallel to Milnor's definition in the non-equivariant case. This definition will be given in §4; we then show in §5 that the formula obtained for L(X/G) in the differentiable case holds more generally when X is a rational homology G-manifold, and indeed can be generalised to a formula for L(h',X/G) where h' belongs to a finite group of automorphisms of X/G induced by automorphisms of X.

A more precise statement of the results proved is given in §1.

The following conventions will apply throughout: the word "manifold" will always refer to a connected, closed (= compact and without boundary) manifold, differentiable unless preceded by the words "rational homology." The coefficients for homology and cohomology will always be one of the fields $\mathbb{Q}$, $\mathbb{R}$, or $\mathbb{C}$ of characteristic zero or else a twisted coefficient system locally isomorphic to one of these; thus there will never be any torsion. We will omit notations for the coefficient homomorphisms, so that, for example, we will multiply the class $L(X) \in H^*(X;\mathbb{Q})$ with elements of $H^*(X;\mathbb{C})$ without explicit comment. Cup products will usually be denoted by juxtaposition but sometimes written out as $x \cup y$. Evaluation of a cohomology class on a homology class will be indicated by juxtaposition or by $\langle\ ,\ \rangle$. A class in the cohomology of a disconnected fixed-point set X^g is a class in the cohomology of each component, and expressions like $L'(g,X)[X^g]$ are to be interpreted as sums over the connectedness components of the corresponding cohomology classes evaluated on the fundamental class of the component in question.

§1. Summary of results

Let X be an oriented closed manifold on which a finite group G acts by orientation-preserving diffeomorphisms. It is known that the signature of the quotient space X/G is the average over G of the equivariant signatures Sign(g,X) (of which a precise definition will be given in §2). The complex numbers Sign(g,X) can in turn be calculated from the G-signature theorem of Atiyah and Singer, which states that

$$\mathrm{Sign}(g,X) = \langle L'(g,X),[X^g]\rangle , \qquad (1)$$

where X^g is the submanifold of X consisting of points left fixed by g and $L'(g,X) \in H^*(X^g;\mathbb{C})$ is a certain cohomology class, explicitly given in terms of the characteristic classes of X^g and its equivariant normal bundle in X. If X^g is not orientable, then both L'(g,X) and the fundamental class are to be understood with the appropriate twisted coefficients; this will be made more precise in §2.

We also know that X/G is a rational homology manifold, and therefore, by the work of Thom and Milnor, has a rational Pontrjagin class and a rational L-class. Since this class is determined by the signatures of the various submanifolds (or rather rational homology submanifolds), it is reasonable to assume that the L-class of X/G can be calculated as was the signature, namely by averaging over G some equivariant L-class in $H^*(X)$. To get from the cohomology of X/G to that of X we simply need to apply π^*, where π is the projection map from X to X/G. We therefore can reasonably expect a formula of the form

$$\pi^*L(X/G) = \frac{1}{|G|} \sum_{g\in G} L(g,X)$$

to hold, where L(g,X) is a class in $H^*(X)$ defined solely by the action of g on X. For the definition of L(g,X), we observe that the G-signature theorem already provides us with a class in $H^*(X^g)$. The map j^* induced by the inclusion j of X^g in X goes the wrong way to define L(g,X) from L'(g,X), so we use instead the "Umkehr homomorphism" or Gysin homomorphism $j_!: H^*(X^g) \to H^*(X)$, defined by passing to homology via Poincaré duality and then applying j_* in homology (we will define all of these concepts more precisely in §2). Finally, to have (1) hold, we need to insert a factor (deg π) to compensate for the difference between the classes $\pi_*[X]$ and $[X/G]$ in $H_*(X/G)$. Thus the formula which we would expect to hold, and which will be proved in §3, is:

Theorem 1: Let G be a finite group, and X an orientable, closed, differentiable G-manifold. Let $\pi: X \to X/G$ denote the projection map, $j: X^g \subset X$ the inclusion of the fixed-point set of an element $g \in G$, and $L'(g,X) \in H^*(X^g)$ the Atiyah-Singer class. Then

$$\frac{1}{\deg \pi} \quad \pi^* L(X/G) \quad = \quad \frac{1}{|G|} \quad \sum_{g \in G} L(g,X), \tag{2}$$

where

$$L(g,X) \quad = \quad j_! \; L'(g,X) \quad \in \quad H^*(X). \tag{3}$$

We make a few comments about the statement of the theorem. The class $L'(g,X)$, as stated above, may lie in $H^*(X^g; \widetilde{\mathbb{C}})$ with $\widetilde{\mathbb{C}}$ a twisted coefficient system locally isomorphic to $\mathbb{C}$ ($\widetilde{\mathbb{C}}$ is the tensor product of the orientation bundle of X^g with the trivial bundle with fibre $\mathbb{C}$), but from the definition of $j_!$ it follows that the class $L(g,X)$ defined by (3) is an untwisted class, so that the summands on the right-hand side of (2) are elements of $H^*(X;\mathbb{C})$. We then deduce from (2) that the sum lies in (the image in $H^*(X;\mathbb{C})$ of) $H^*(X;\mathbb{Q})$. Thus, even if we are not interested in the quotient X/G or its L-class as such, we still get interesting information about the G-space X itself from equation (2), namely a sort of integrality theorem for the cohomology classes defined by (3). The Atiyah-Singer result only gives the top-dimensional component of this (i.e. the signatures), but then gives a stronger result: the average over G of the complex (algebraic) numbers Sign(g,X) is not only a rational number, but even a rational integer.

The next point about formula (2) is that if G acts effectively (which can always be assumed by factoring out the normal subgroup which acts trivially), then $\deg \pi = |G|$, so the numerical coefficients can be omitted from (2), simplifying it somewhat.

Finally, it is known that, for cohomology with rational or complex coefficients (or, more generally, any field of characteristic zero or prime to $|G|$ as coefficients) the map π^* induces an isomorphism

$$\pi^* : \; H^*(X/G) \longrightarrow H^*(X)^G \subset H^*(X) \tag{4}$$

from $H^*(X/G)$ onto the G-invariant part of $H^*(X)$ (Grothendieck [10],

Borel [2]). In particular π^* is injective, so eq. (2) determines the L-class of X/G completely.

Theorem 1 as stated above is the first main result of this chapter. However, its proof suggests the possibility of defining the cohomology classes $L(g,X) \in H^*(X;\mathbb{C})$ when X is just an oriented rational homology G-manifold. Here we cannot use a formula such as (3), since it is not possible to define the Atiyah-Singer class for non-differentiable actions, and indeed it is not clear that the class L(g,X) that we define vanishes if g acts freely on X. But we can still define the class, in a way exactly parallel to the Milnor definition of L(X) for a rational homology manifold X. This will be done in §4. Once L(g,X) is defined, eq. (2) makes sense even for rational homology manifolds X (since X/G is then also a rational homology manifold and has an L-class), and we prove in §5 that it still holds. Indeed, we can generalise it:

Theorem 2: Let X be an oriented rational homology G-manifold. For each g in G, let L(g,X) in $H^*(X;\mathbb{C})$ be the class defined in §4. If h is an automorphism of X of finite order which commutes with the action of G on X, and h' the induced automorphism of X/G, we have the relation

$$\frac{1}{\deg \pi} \pi^* L(h',X/G) = \frac{1}{|G|} \sum_{g \in G} L(gh,X). \qquad (5)$$

Finally, in §6 we evaluate explicitly the quantities L(g,X) for X = complex projective space and G acting linearly on X. This calculation will be used in §15 to check a general formula for L(g,X) when X is the nth symmetric product of a manifold (the space $P_n\mathbb{C}$ is the nth symmetric product of S^2). When we put the value of L(g,X) into Theorem 1 we obtain a formula for L(X/G) already obtained by Bott by other methods (unpublished; see, however, Hirzebruch [16]).

§2. Preparatory material

This section contains more detailed descriptions of some of the concepts and theorems which were used in §1 for the formulation of the various results stated there. We do not discuss definitions or results which are very well known. Thus, for example, we assume the definitions of the signature and the L-class of a manifold and a knowledge of the Hirzebruch index theorem, but define the equivariant versions of these notions and state explicitly the G-signature theorem of Atiyah and Singer. We also define rational homology manifolds and give in some detail Milnor's formulation of the definition of the L-class for a rational homology manifold (that a definition is possible had been shown by Thom). This will be especially important to us since we will copy the construction in §4 for the definition of the equivariant L-class $L(g,X)$ for rational homology manifolds X. The only point we need to make for a reader acquainted with these ideas and wishing to skip this section is that the class $L'(g,X)$ appearing in §1 is not exactly the cohomology class appearing in the original formulation of the G-signature theorem (Atiyah and Singer [1]) but differs from it by a power of two in each dimension (except the component of top degree, which for both classes is equal to $\mathrm{Sign}(g,X)$).

We break up the section into three parts. In (I) we discuss various homological properties of manifolds: the definition of a rational homology manifold, the orientation system of local coefficients for a manifold, and related concepts (the Thom class of a non-oriented bundle, Poincaré duality for a non-orientable manifold, the Gysin homomorphism). A description of Milnor's definition of the L-class of a rational homology manifold then follows in (II), while (III) contains the definition of $\mathrm{Sign}(g,X)$ for a G-manifold X and a statement of the Atiyah-Singer G-signature theorem.

(I) Homological properties of manifolds

A rational homology manifold of dimension $\underline{n}$ is a triangulated space in which the boundary of the star of each vertex has the same rational homology groups as S^{n-1}. Equivalently, it is a simplicial complex X such that

$$H^i(X, X-\{x\};\mathbb{Q}) \approx H^i(\mathbb{R}^n, \mathbb{R}^n-\{0\};\mathbb{Q}) \tag{1}$$

for all $x \in X$. Then we can define a system of local coefficients for X, denoted or_X and called the <u>orientation</u> <u>system</u> of X, which at the point <u>x</u> is just the vector space $H^n(X,X-\{x\};\mathbb{Q})$ (where we do not choose a specific isomorphism with $\mathbb{Q}$). There is then an <u>orientation</u> <u>class</u>

$$[X] \in H_n(X;or_X) \tag{2}$$

whose image in $H_*(X,X-\{x\})$ for $x \in X$ is the identity in

$$Hom(H^n(X,X-\{x\}),\ H^n(X,X-\{x\})) \cong H_n(X,X-\{x\},or_X|_x). \tag{3}$$

Notice that this is independent of the particular isomorphisms of $H^n(X,X-\{x\})$ with $\mathbb{Q}$ given by (1). If the system or_X of local coefficients is trivial, X is said to be <u>orientable</u>; then the element (2) is in $H_n(X;\mathbb{Q})$.

If Γ is any system of local coefficients for X, then the cap product with $[X]$ gives a <u>Poincaré</u> <u>duality</u> <u>isomorphism</u>:

$$\cap [X] = D_X : H^i(M;\Gamma) \xrightarrow{\cong} H_{n-i}(M;\Gamma \otimes or_X). \tag{4}$$

If $f:X\to Y$ is a map between two manifolds, the <u>Gysin</u> <u>homomorphism</u> $f_!$ is

$$f_! = D_Y^{-1} f_* D_X : H^*(X;f^*\Gamma \otimes or_X) \longrightarrow H^*(Y;\Gamma \otimes or_Y), \tag{5}$$

where Γ is a local coefficient system over Y, $f^*\Gamma$ the induced system over X, and f_* the map from $H_*(X;f^*\Gamma)$ to $H_*(Y;\Gamma)$. We will only need this when <u>f</u> is an inclusion map between differentiable manifolds and $\Gamma = \mathbb{Q}$.

If ξ is a real vector bundle of dimension <u>q</u> over X, it also defines an orientation system of local coefficients, which at $x \in X$ is

$$or_\xi(x) = H_q(E_x,E_x-\{0\}), \tag{6}$$

the fibre of ξ at <u>x</u> being denoted E_x. This defines a <u>Thom</u> <u>class</u>

$$U_\xi \in H^q(E.E_\bullet;\pi^* or_\xi), \tag{7}$$

where $E_\bullet$ is the total space E of ξ minus the zero-section X, π is the projection map $E\to X$, and $\pi^* or_\xi$ is the system of local coefficients on E induced from or_ξ by π. Namely, the Thom class is defined uniquely by the requirement that its restriction to any fibre is the identity of

$$H^q(E_x,E_{\bullet x};\pi^* or_\xi) \cong Hom(H_q(E_x,E_{\bullet x}),\ H_q(E_x,E_{\bullet x})). \tag{8}$$

Then the Euler class of ξ is defined as the restriction to the zero-section X of the Thom class; thus

$$e(\xi) \quad \epsilon \quad H^q(X;or_\xi). \tag{9}$$

The bundle ξ is oriented if or_ξ is trivial; then its Euler class lies in $H^*(X;\mathbb{Z})$ (or $H^*(X;\mathbb{Q})$ if we used rational coefficients in (6)). This is the case if and only if the first Stiefel-Whitney class of ξ is zero. In general, the first Stiefel-Whitney class determines the orientation system of coefficients; thus when we take a direct sum $\xi \oplus \eta$ of bundles, the orientation system of the sum is the tensor product $or_\xi \otimes or_\eta$, while the Stiefel-Whitney class of the sum is the sum $w_1(\xi) + w_1(\eta)$. In particular, since Stiefel-Whitney classes have order two, it does not matter in equations like (4) whether we tensor with or_X or $or_X{}^{-1}$ (we have already used this freedom in eq. (5), where the positions of Γ and $\Gamma \otimes or_X$ have been interchanged). Finally, if X is a differentiable manifold, then or_X equals or_{TX}, where TX is the tangent bundle.

(II) Milnor's definition of the L-class of a rational homology manifold

It has been proved by Serre [17] that, if X is a CW complex of dimension $n \leq 2i-2$, then homotopy classes of maps

$$f: X \rightarrow S^i \tag{10}$$

form an abelian group $\pi^i(X)$ (the ith cohomotopy group of X) which, up to torsion, is isomorphic to $H^i(X)$. More precisely, the natural map $\pi^i(X) \rightarrow H^i(X)$ sending the map (10) to $f^*\sigma$ (where $\sigma \in H^i(S^i)$ is the generator) becomes an isomorphism after tensoring with $\mathbb{Q}$. Thus, **for any element $x \in H^i(X)$, some multiple Nx can be written as $f^*\sigma$ for some $f:X \rightarrow S^i$ whose homotopy type is unique up to torsion in $\pi^i(X)$.**

Now let X be an oriented rational homology manifold of dimension n. Then there is a fundamental class [X] in $H_n(X;\mathbb{Q})$, and therefore we have an intersection form on $H^k(X)$ (if n=2k) defined as usual by sending two elements to the evaluation of their cup product on [X]; thus the signature of X is defined. Now one can prove the following facts: if f is a simplicial map as in (10), where S^i has some fixed standard triangulation, then the inverse image of a point,

$$A = f^{-1}(p) \subset X, \tag{11}$$

is an oriented rational homology submanifold of X (of dimension n-i) for almost all $p \in S^i$, and moreover, the cobordism class of A--and hence also its signature--are independent of the point p, again for almost all p. This defines then a number I(f) = Sign(A), which only depends on the homotopy type of A. Moreover, from the definition of the addition in $\pi^i(X)$ one easily finds $I(f_1+f_2) = I(f_1) + I(f_2)$, so the map $[f] \to I(f)$ defines a homomorphism

$$\pi^i(X) \xrightarrow{I} \mathbb{Z}. \tag{12}$$

If we tensor this with $\mathbb{Q}$ and combine it with the theorem of Serre stated above and the Poincaré duality isomorphism in X, we obtain a unique class

$$l_{n-i} \in H^{n-i}(X;\mathbb{Q}) \tag{13}$$

such that

$$(l_{n-i} \cup f^*\sigma)[X] = I(f) = \mathrm{Sign}(A) \tag{14}$$

for all maps f as in (10). This all only holds for $n \leq 2i-2$ or $n-i \leq (n-2)/2$, but we can define l_j also for larger j by choosing a large integer N, defining the class $l_j' \in H^j(X\times S^N)$ by the above procedure applied to $X\times S^N$, and then letting l_j be the corresponding class under the isomorphism of $H^j(X)$ with $H^j(X\times S^N)$. Then the L-class of X is defined as the sum of these classes:

$$L(X) = \sum_{j=0}^{\infty} l_j \in H^*(X;\mathbb{Q}). \tag{15}$$

It is easy to see that this agrees with the usual definition if X is a differentiable manifold. Indeed, since the L-class of a product is multiplicative and $L(S^N)=1$, we have $L(X) = L(X\times S^N)$ (where we have identified the cohomologies of the two spaces in dimensions up to $n \ll N$), so we only have to check that the usual L-class satisfies (14). But the map f can be chosen within its homotopy class as differentiable; then A is a differentiable submanifold of X for almost all p in S^i and has trivial normal bundle (since a point has trivial normal bundle), so the L-class of A is the restriction $j^*L(X)$ of the L-class of X (where $j: A \subset X$). Also, $f^*\sigma \in H^i(X)$ is the Poincaré dual of $j_*[A] \in H_{n-i}(X)$. Therefore

$$\langle L(X) \cup f^*\sigma, [X]\rangle = \langle L(X), j_*[A]\rangle = \langle j^*L(X), [A]\rangle = \langle L(A), [A]\rangle = \mathrm{Sign}(A),$$

so the L-class in the usual sense satisfies (14). Since equation (14) defined uniquely the class l_j (for X a rational homology manifold), we see that the class (15) is indeed the usual L(X) if X is differentiable.

(III) The G-signature theorem

If a group G acts on an orientable rational homology manifold X (here and in future we will tacitly assume that all groups are compact Lie groups and that actions preserve any structure present; thus the action here must be simplicial and preserve the orientation), there is defined a complex number Sign(g,X), depending only on the action of g^* on $H^*(X;\mathbb{R})$, for every $g \in G$. The definition is as follows: If X has odd dimension, we set Sign(g,X) = 0. If X has dimension 4k, the intersection form $B(x,y) = \langle x \cup y, [X] \rangle$ is symmetric, non-degenerate (because of Poincaré duality), and G-invariant (since $g_*[X] = [X]$ by assumption). We can therefore decompose the middle cohomology group $H^{2k}(X;\mathbb{R})$ in a G-invariant way as a direct sum $H_+ \oplus H_-$, where the bilinear form B is positive definite on H_+ and negative definite on H_-. Then

$$\mathrm{Sign}(g,X) = \mathrm{tr}(g^*|H_+) - \mathrm{tr}(g^*|H_-); \qquad (16)$$

this definition is independent of the decomposition $H^{2k}(X) = H_+ \oplus H_-$. If X has dimension 4k+2, then B(x,y) is skew-symmetric, non-degenerate and G-invariant. Then if we choose a G-invariant positive definite inner product $\langle\,,\,\rangle$ on $H^{2k+1}(X;\mathbb{R})$ and define an operator A by requiring $B(x,y) = \langle Ax,y\rangle$ for all $x,y \in H^{2k+1}(X;\mathbb{R})$, the operator A is skew-adjoint, so the operator $J = A/(AA^*)^{1/2}$ has square -1. This gives $H^{2k+1}(X;\mathbb{R})$ the structure of a complex vector space, and since g^* commutes with J, it acts on $H^{2k+1}(X;\mathbb{R})$ in a complex linear way. Then

$$\mathrm{Sign}(g,X) = 2i\,\mathrm{Im}\,(\,\mathrm{tr}(g^*|H^{2k+1}(X;\mathbb{R}))), \qquad (17)$$

where the trace is taken of the map g^* thought of as an automorphism of a complex vector space. Again this is independent of the choices made. This completes the definition of Sign(g,X) in all cases. If g_1 acts on X_1 and g_2 acts on X_2, then the signature of the product action of $g_1 \times g_2$ on $X_1 \times X_2$ is given by

$$\mathrm{Sign}(g_1\times g_2, X_1\times X_2) = \mathrm{Sign}(g_1,X_1)\cdot\mathrm{Sign}(g_2,X_2). \qquad (18)$$

The content of the G-signature theorem is a formula for Sign(g,X) in terms of the fixed-point set X^g and its equivariant normal bundle, in the case that X is a differentiable manifold. To state this formula, we first define certain characteristic classes of complex and real bundles (i.e. multiplicative sequences in the Chern or Pontrjagin classes, respectively). If θ is a real number, not a multiple of π, and ξ is a complex bundle of (complex) dimension q over a space Y, we define

$$L_\theta(\xi) = (\coth \frac{i\theta}{2})^q \prod_j \frac{\coth(x_j + i\theta/2)}{\coth i\theta/2}, \tag{19}$$

where the x_j have the usual interpretation as formal two-dimensional cohomology classes such that the Chern class of ξ is

$$c(\xi) = \prod_j (1 + x_j). \tag{20}$$

Thus $L_\theta(\xi)$ is an element of the subring $H^*(Y;\mathbb{Q})[e^{i\theta}]$ of $H^*(Y;\mathbb{C})$. If ξ is a real bundle over Y, we let

$$L(\xi) \in H^*(Y;\mathbb{Q}) \tag{21}$$

be the Hirzebruch L-class of ξ, defined as $\prod \frac{x_j}{\tanh x_j}$ where $\prod(1+x_j^2)$ is the Pontrjagin class of ξ. We let

$$e(\xi) \in H^*(Y;or_\xi) \tag{22}$$

be the Euler class (cf. eq. (9) above), and define

$$L_\pi(\xi) = e(\xi)L(\xi)^{-1} \in H^*(X;or_\xi \otimes \mathbb{Q}), \tag{23}$$

which is legitimate since $L(\xi)$ has leading coefficient 1 and is therefore invertible. Notice that if ξ is a complex bundle and we set $\theta=\pi$ in (19) (where we have chosen the number of x_j's to be equal to $\underline{q}$) we obtain, after first cancelling the (zero!) factors $\coth i\theta/2$ from numerator and denominator, a produce of the $\tanh x_j$. This then agrees with (23), since $e(\xi) = c_q(\xi) = x_1 \dots x_q$ and $L(\xi) = \prod (x_j/\tanh x_j)$ in this case.

We return to the action of $\underline{g}$ on a differentiable manifold X. The fixed-point set X^g is a smooth submanifold, not necessarily orientable; we denote its normal bundle in X by N^g. At each point $\underline{x}$ of X, the action of $\underline{g}$ on the fibre N^g_x can be decomposed, by standard representation theory, as a sum of one-dimensional subspaces on which g acts as multiplication by -1 and two-dimensional subspaces on which g acts by

$$A_\theta = \begin{pmatrix} \cos\theta & -\sin\theta \\ \sin\theta & \cos\theta \end{pmatrix}, \tag{24}$$

where θ is a real number not divisible by π (the eigenvalue +1 cannot occur on the normal bundle N^g). Since A_θ and $A_{-\theta}$ are equivalent, we can assume that $0<\theta<\pi$ for the representations (24), and write

$$N^g_x = N^g_{x,\pi} \oplus \sum_{0<\theta<\pi} N^g_{x,\theta}, \tag{25}$$

where $N^g_{x,\pi}$ is the subspace on which g acts as -1. Each of the spaces $N^g_{x,\theta}$ $(0<\theta<\pi)$ has a natural complex structure on which g acts as multiplication by $e^{i\theta}$. The decomposition (25) of a fibre extends to a decomposition of the whole bundle N^g as

$$N^g = N^g_\pi \oplus \sum_{0<\theta<\pi} N^g_\theta, \tag{26}$$

where N^g_π is now a real bundle over X^g on which g acts as -1 and N^g_θ is a complex bundle over X^g on which g acts as $e^{i\theta}$. In particular, the bundles N^g_θ all acquire a natural orientation from the complex structure. Since X also has a given orientation, we obtain from (26) (and the relation $j^*(TX) = N^g \oplus TX^g$) an isomorphism between the systems of twisted coefficients or_{X^g} and $or_{N^g_\pi}$ (cf. (I) of this section). We now define a class

$$L'(g,X) = L(X^g)\cdot L_\pi(N^g_\pi)\cdot \prod_{0<\theta<\pi} L_\theta(N^g_\theta) \in H^*(X^g; or_{X^g}\otimes\mathbb{C}), \tag{27}$$

where we have used the characteristic classes defined above and the isomorphism $or_{X^g} \approx or_{N^g_\pi}$ given by the prescribed orientations on X and on the complex bundles N^g_θ. Finally, the fundamental class $[X^g]$ also lies in the homology group with coefficients or_{X^g}, so we can evaluate (27) on this class. The Atiyah-Singer G-signature theorem states that this is precisely Sign(g,X):

$$\mathrm{Sign}(g,X) = \langle L'(g,X), [X^g]\rangle. \tag{28}$$

<u>Sources</u>: The material in (I) is standard; it can be found sketchily in Spanier [38] or Dold [8] and in detail in Heithecker [11]. The original

paper of Thom proving that L-classes can be defined for rational homology manifolds in a way consistent with the previous definition in the differentiable case is Thom [39]. The definition of rational homology manifolds as given in (I) and the whole contents of (II) are taken from Milnor [31]. Finally, the material in (III) comes from Atiyah and Singer [1].

§3. Proof of the formula for L(X/G)

The starting point for the proof of Theorem 1 of §1 will be the corresponding relation for signatures, namely

$$\mathrm{Sign}(A/G) = \frac{1}{|G|} \sum_{g \in G} \mathrm{Sign}(g,A) \tag{1}$$

for a G-manifold A (G finite). In fact this also holds for a rational homology manifold A, since its proof depends only on the definition of Sign(g,A) in terms of the action of g^* on $H^*(A)$. It is trivial if A has odd dimension (both sides are zero) or dimension 4k+2 (then the left-hand side is identically zero, while $\mathrm{Sign}(g,A) = -\mathrm{Sign}(g^{-1},A)$). If the dimension of A is 4k, we use the fact (cf. §1) that

$$\pi^*: H^*(A/G) \xrightarrow{\cong} H^*(A)^G , \tag{2}$$

where π denotes the projection map from A to A/G (this holds because we are always working with a field of characteristic zero as coefficients). Then comparing the definition of Sign(g,A) (§2, (16)) in this case with the definition of Sign(A/G), we find that (1) reduces to the elementary identity of linear algebra

$$\dim V^G = \frac{1}{|G|} \sum_{g \in G} \mathrm{tr}\,(g|V) \tag{3}$$

for the G-invariant part of a vector space V.

We will apply this to the set $A = f^{-1}(p)$ appearing in Milnor's definition of L-classes (§2, (11)). Since we are interested in the L-class of X/G, we begin by choosing a simplicial map

$$\bar{f}: X/G \to S^i . \tag{4}$$

We then define an equivariant map $\underline{f}$ from X to S^i by

$$f = \bar{f} \circ \pi : X \to S^i \tag{5}$$

Clearly, $\underline{f}$ is G-equivariant with G acting trivially on S^i, i.e. it is G-invariant . We write

$$A = f^{-1}(p) \subset X, \tag{6}$$

$$\bar{A} = \bar{f}^{-1}(p) = A/G \subset X/G. \tag{7}$$

Thus A is a G-invariant subspace of X.

We can change f within its equivariant homotopy class (i.e. through maps factoring through X/G) to a differentiable map. Indeed, we just replace the composition $f: X \to S^i \to \mathbb{R}^{i+1}$ by a differentiable map $f': X \to \mathbb{R}^{i+1}$ with $\max |f(x) - f'(x)| < \varepsilon$, and then define a third map f" from X to $\mathbb{R}^{i+1}$ by

$$f''(x) = \frac{1}{|G|} \sum_{g \in G} f'(g \circ x). \tag{8}$$

Then f" is G-equivariant and differentiable and, if ε is small enough, close to f; in particular $f''(X) \subset \mathbb{R}^{i+1} - \{0\}$ so we can compose with the projection $\mathbb{R}^{i+1} - \{0\} \to S^i$ to obtain a map $f''': X \to S^i$ which is G-equivariant, differentiable, and arbitrarily close to f. If we apply the whole process to f_t (where f_t is a homotopy from f to f', e.g. the linear one) we obtain an equivariant homotopy from f to f‴.

We therefore assume that f is differentiable. For the whole of Milnor's definition we are allowed to exclude sets of measure zero for the point p in S^i; thus here we can use Sard's theorem to have p a regular value of the function f and of each of the (finitely many) functions $f|X^g$ ($g \in G$; notice that X^g is a differentiable manifold). Then A is a differentiable submanifold of X and meets each submanifold X^g transversally with intersection A^g.

We now write L for L(X/G) for convenience; then Milnor's definition says that L is uniquely determined by

$$(L \cup \bar{f}^*\sigma)[X/G] = \operatorname{Sign}(\bar{A}) \tag{9}$$

(again for almost all $p \in S^i$; we are also ignoring the problem of defining the classes l_j for j larger than half the dimension of X since we can remove such dimensional restrictions by multiplying everything with a sphere of large dimension on which G acts trivially).

We now apply (1) to express the right-hand side of (9) in terms of the quantities Sign(g,A), and then the G-signature theorem (§2 (28)) to evaluate the latter. Thus

$$(L \cup \bar{f}^*\sigma)[X/G] = \operatorname{Sign}(A/G) = \frac{1}{|G|} \sum_{g \in G} \operatorname{Sign}(g,A)$$

$$= \frac{1}{|G|} \sum_{g \in G} L'(g,A)[A^g]. \tag{10}$$

The G-signature theorem is applicable because A is a differentiable manifold on which G acts.

In the diagram

$$\begin{array}{ccc} A^g = A \cap X^g & \xrightarrow{\ i'\ } & X^g \\ {\scriptstyle j'}\downarrow & & \downarrow{\scriptstyle j} \\ A & \xrightarrow{\ i\ } & X \end{array}$$

of differentiable manifolds and inclusion maps, the inclusions i and i' consist of the inclusion of the inverse image of a point (of the differentiable maps $\underline{f}$ and $f|X^g$, respectively) and therefore have trivial normal bundle. It follows that

$$N(A^g) \oplus \varepsilon \quad \approx \quad i'^* N(X^g) \oplus \varepsilon \tag{11}$$

where $N(A^g)$ denotes the normal bundle of A^g in A, $N(X^g)$ denotes that of X^g in X, and ε denotes a trivial bundle (of dimension $\underline{i}$). This isomorphism is even G-equivariant since $A = f^{-1}(p)$ and $\underline{f}$ is a G-invariant map. Therefore the eigenvalues of the action of $\underline{g}$ on $N(A^g)$ are the same as those of its action on $N(X^g)$, and the corresponding eigenbundles N^g_θ also correspond under i'^* as in (11). We therefore deduce from the definition of L' (§2 (27)) that

$$L'(g,A) \quad = \quad i'^* L'(g,X). \tag{12}$$

If we put this into (10) and calculate as in the non-equivariant case (§2), we find:

$$\begin{aligned} < L'(g,A),[A^g] > \quad &= \; < i'^* L'(g,X),[A^g] > \\ &= \; < L'(g,X), i'_*[A^g] > \\ &= \; < L'(g,X),\ D_{X^g}[(f|X^g)^*\sigma] > \end{aligned}$$

(here as in §2 we use that the homology class $[f^{-1}(p)]$ for a map of the form of §2 (10) is the Poincaré dual of the class $f^*\sigma$, where σ is the generator of $H^i(S^i)$)

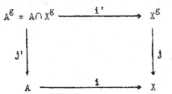

$$= \langle L'(g,X), [(f\circ j)^*\sigma] \cap [X^g]\rangle$$

(from the definition of Poincaré duality)

$$= \langle L'(g,X) \cup j^*f^*\sigma, [X^g]\rangle$$

$$= \langle j^*f^*\sigma \cup L'(g,X), [X^g]\rangle$$

(one can interchange since $L'(g,X)$ and X^g are even- or odd-dimensional classes according as X is, and therefore have the same parity, so the expression is zero unless $i = \deg\sigma = \deg j^*f^*\sigma$ is even)

$$= \langle j^*f^*\sigma, L'(g,X) \cap [X^g]\rangle$$

$$= \langle j^*f^*\sigma, D_{X^g}(L'(g,X))\rangle$$

$$= \langle f^*\sigma, j_* D_{X^g}(L'(g,X))\rangle$$

$$= \langle f^*\sigma, D_X(j_! L'(g,X))\rangle$$

$$= \langle f^*\sigma, L(g,X) \cap [X]\rangle$$

$$= \langle L(g,X) \cup f^*\sigma, [X]\rangle \quad , \tag{13}$$

where in the last lines we have substituted the definitions of $j_!$ and of $L(g,X)$ given in §2 and §1, respectively. In this calculation we have not specified the coefficients, but if one follows through the steps with the definitions from §2, part (I) in mind, one finds that the calculation is consistent also when twisted coefficients must be used (i.e. when X^g and hence A^g is non-orientable). In the last line all the cohomology and homology classes appearing have simple coefficients.

We now substitute (13) into (10), obtaining

$$\langle L \cup \bar{f}^*\sigma, [X/G]\rangle = \frac{1}{|G|} \sum_{g\in G} \langle L(g,X) \cup f^*\sigma, [X]\rangle. \tag{14}$$

If we further substitute $[X/G] = \frac{1}{\deg\pi}\, \pi_*[X]$ and $\pi^*\bar{f}^* = f^*$, we get

$$\langle \left(\frac{1}{\deg\pi}\, \pi^*L\right) \cup f^*\sigma, [X]\rangle = \langle \left(\frac{1}{|G|} \sum_{g\in G} L(g,X)\right) \cup f^*\sigma, [X]\rangle, \tag{15}$$

and the desired equality (eq. (2) of §1) follows from this equation and the fact that (9) defines the L-class $L = L(X/G)$ uniquely.

It is interesting to look more closely at this equation and see the relation between the properties of the equivariant classes $L(g,X)$ and the ordinary L-class. These properties of $L(g,X)$ will be used later. They also serve to make Theorem 1 of §1 more plausible by showing that §1(2) defines uniquely a class $L(X/G) \in H^*(X/G)$ which has the properties expected of an L-class (leading coefficient 1, zero in dimensions $\neq 4k$, etc.). The properties in question are:

i) If $g,h \in G$, then

$$h^*L(g,X) = L(h^{-1}gh,X). \tag{16}$$

ii) If G acts effectively and G is connected, then the component of $L(g,X)$ in $H^0(X)$ is 0 if $g\neq 1$, 1 if $g=1$.

iii) For all $g \in G$, the component of $L(g,X) + L(g^{-1},X)$ in $H^i(X)$ is zero unless $\underline{i}$ is divisible by four.

iv) $$L(g,X)[X] = \mathrm{Sign}(g,X). \tag{17}$$

v) The sum $\Sigma_{g\in G} L(g,X)$ is in $H^*(X;\mathbb{Q}) \subset H^*(X;\mathbb{C})$.

It follows from i) that the average over G of the classes $L(g,X)$ is invariant under the action of G on $H^*(X)$, and therefore (by the isomorphism (4) of §1) that it is π^* of a unique element of $H^*(X/G)$. If we write this element as $L/\deg \pi$, then it follows from ii) that L has leading coefficient 1, from iii) that L is non-zero only in dimensions 4k, from iv) that $L[X] = \mathrm{Sign}(X)$, and from v) that L is a rational cohomology class. Thus L has all the properties required if it is to be equal to $L(X/G)$.

The proofs of i)-iv) are quite simple. Property iv) follows from the G-signature theorem (§2(28)) and the definition of $j_!$. Property ii) is clear since the map $j_!$ raises dimensions by the difference of the dimensions of its domain and target manifolds, so $j_!L'(g,X)$ can have a zero-dimensional component only if $\dim X^g = X$, which for X connected can only occur if $X^g=X$ and therefore (since the action is effective) if $g=1$. To prove i), we observe that the map $h:X\to X$ defined by the action of $h \in G$ maps X^g isomorphically onto $X^{g'}$ ($g'=hgh^{-1}$) and that the map h^* pulls back the normal bundle $N^{g'}$ to N^g (as G-bundles, i.e. the splitting into eigenbundles is also pulled back). It follows from this and from the functoriality of the characteristic classes L, L_π and L_θ appearing in the definition of $L'(g,X)$ that

$$L'(g,X) = h^*L'(hgh^{-1},X). \tag{18}$$

Moreover, since $\underline{h}$ is an isomorphism we have $h^* = (h^{-1})_!$, and therefore equation (16) follows from equation (18). To prove iii), we note that the elements g and g^{-1} have the same fixed-point sets, and that the eigenvalue decompositions of the common normal bundle N^g are related by $N^g_\theta \cong N^{g^{-1}}_{-\theta}$. In particular $N^g_\pi \cong N^{g^{-1}}_\pi$. We substitute this into the definition §2(27) of $L'(g,X)$, using the fact that $j_!$ increases dimensions by $\dim X - \dim X^g$ and the fact that $L_\pi(N^g_\pi) = e(N^g_\pi)L(N^g_\pi)^{-1}$ only has components in dimensions $\equiv \dim N^g_\pi \pmod 4$. We find that the proof of iii) reduces to showing that

$$\prod_{0<\theta<\pi} L_\theta(N^g_\theta) + \prod_{0<\theta<\pi} L_{-\theta}(N^g_\theta) \in H^*(X^g;\mathbb{C})$$

only has components in degrees equal (modulo 4) to

$$\dim X - \dim X^g - \dim N^g_\pi = 2 \sum_{0<\theta<\pi} \dim_{\mathbb{C}} N^g_\theta .$$

But this follows easily from the identity $\coth(x_j - i\theta) = -\coth(-x_j + i\theta)$. We will not give a complete proof of property v) here, since in any case it follows from Theorem 1 of §1 and the rationality of $L(X/G)$. The method of proof is to write

$$\sum_{g\in G} L(g,X) = \sum_{Y\subset X} L_Y, \tag{19}$$

where, for Y a connected closed submanifold of X, L_Y denotes the sum over G of the contribution from Y to $L(g,X)$ (i.e. zero unless Y is a component of X^g, and $j_![L(Y)\prod_\theta L_\theta(N^g_\theta)]$ if Y is such a component, where j is the inclusion of Y in X and N^g_θ is the $e^{i\theta}$ eigenbundle of the action of g on the normal bundle of Y in X). This makes sense, since L_Y is zero for all but finitely many submanifolds Y. One then can show that each of the classes $L_Y \in H^*(X)$ is a rational cohomology class by a Galois-theory type of argument. The argument when Y is a single point $\{x\}$ is given in the introduction to Chapter III.

§4. A definition of $L(g,X)$ for rational homology manifolds

The reason that the L-class can be defined for a rational homology manifold X is that the L-class is related to the signature of certain submanifolds of X, and that there are enough of these submanifolds to determine $L(X)$ completely. It is reasonable to ask whether the equivariant L-class has similar properties which allow its definition for rational homology G-manifolds.

We cannot expect such a definition for the Atiyah-Singer class $L'(g,X)$, since it lies in the cohomology of X^g and therefore only can be defined in terms of the local action of g near its fixed-point set, which presupposes that this action is differentiable, or at least that it looks like a differentiable action in a neighbourhood of X^g (cf. Wall [40], Ch. 14). But the class $L(g,X) \in H^*(X)$, defined in the differentiable case as $j_! L'(g,X)$, can be characterised in certain circumstances by a formula proved in the last section in the course of proving Theorem 1 of §1, namely (eq. (13) of §3)

$$\mathrm{Sign}(g,A) \;=\; \langle L(g,X) \cup f^*\sigma, [X] \rangle \,, \tag{1}$$

where $A = f^{-1}(p)$, f being a map from X to S^i which is G-equivariant (G acts trivially on S^i), p a sufficiently general point of S^i, and σ a generator of $H^i(S^i;\mathbb{Z})$. Equivalently, if we use the fact that $f^*\sigma$ is the Poincaré dual in X of the homology class $i_*[A]$ (this was used in the proof given in §3), we can consider (1) as saying that the value of $L(g,X)$ on a given homology class is $\mathrm{Sign}(g,A)$, where A is a G-invariant submanifold of X which represents this class and which is the inverse image of a point for some map $X \to S^i$ (in the differentiable case, this says that A has trivial normal bundle in X).

We thus wish to define a class

$$L(g,X) \;\in\; H^*(X;\mathbb{C}) \tag{2}$$

for a rational homology G-manifold X in such a way that (1) still holds. For this we require that there are enough "G-invariant submanifolds A with trivial normal bundle" in the sense defined above, i.e. that such manifolds exist in enough homology classes of X to determine $L(g,X)$ completely, and also that there are not too many, so that the conditions (1) do not conflict with one another. We cannot expect that all of the elements of $H_*(X)$ are represented by embedded manifolds A of the type desired (as was the case for Milnor's definition, at least in rational homology), since a G-invariant submanifold A of X can certainly only represent a G-invariant homology class. Therefore (1) only tells us how to evaluate on elements of $H_*(X)^G$, or rather only on those elements of $H_*(X)^G$ which can be represented by good submanifolds A. To have a reasonable hold on the group $H_*(X)^G$, we must assume that G is finite, in which case this group is isomorphic to $H_*(X/G)$ under the map π_* (cf. (4) of §1). Then knowing the value of a cohomology class only on the G-invariant part of $H_*(X)$ only determines the cohomology class if it is itself G-invariant (for then it corresponds to an element of $H^*(X/G)$ and is determined by its values on elements of $H_*(X/G)$). But we saw in §3 that

$$h^*L(g,X) \quad = \quad L(h^{-1}gh,X) \in H^*(X;\mathbb{C}) \qquad\qquad (\text{all } h \in G) \qquad\qquad (3)$$

in the differentiable case, and it is easy to see that the same formula will hold for a class $L(g,X)$ defined using (1) (just replace $\underline{f}$ in eq. (1) by $f \circ h$, which is also a G-invariant map from X to S^1). Therefore if we want the cohomology class $L(g,X)$ to be invariant under the action of G on $H^*(X)$, we should require that $h^{-1}gh = g$ for all $g,h \in G$, i.e. that G be abelian. Unlike the requirement that G be finite, however, this does not limit the generality of our definition, since we want our class $L(g,X)$ to share with the Atiyah-Singer class the property of depending only on $\underline{g}$ and its action on X but not on G, and therefore we can always replace G by the abelian subgroup generated by $\underline{g}$.

We can now state the theorem of this section:

<u>Theorem 1</u>: Let X be an oriented rational homology G-manifold, G finite abelian. There is a unique class $L(g,X)$ in $H^*(X;\mathbb{C})^G$ satisfying (1) for all simplicial G-invariant maps $f:X \to S^1$, and this class agrees with the class $L(g,X)$ of §3 if X and the action of G are differentiable.

<u>Proof</u>: The last statement follows from the assertion about uniqueness, since we saw in §3 that the differentiably defined class $L(g,X)$ satisfies (1).

We now proceed as in the non-equivariant case outlined in §2. We work with complex coefficients, so that we are allowed to use the isomorphism (4) of §1. We also assume that we have multiplied X with a G-invariant sphere of large dimension to remove restrictions on the dimensions of the cohomology groups to which we can apply Serre's theorem. Then $H^*(X/G;\mathbb{C})$ is generated by cohomology classes which can be represented by maps $\bar{f}:X/G \to S^i$. Such maps define G-invariant maps $f:X \to S^i$ by $f=\bar{f}\circ\pi$, and conversely any G-invariant map $\underline{f}$ factors through X/G. The same applies to equivariant homotopies. We thus obtain a commutative diagram in which all arrows are isomorphisms:

$$\begin{array}{ccc} H^i(X/G;\mathbb{C}) & \xrightarrow{\pi^*} & H^i(X;\mathbb{C})^G \\ \downarrow & & \downarrow \\ \pi^i(X/G)\otimes\mathbb{C} & \xrightarrow{\pi^*} & \pi^i(X)^G\otimes\mathbb{C} \end{array} \tag{4}$$

We now define, for each $p \in S^i$, a map

$$T_p:\ \pi^i(X)\otimes\mathbb{C} \to \mathbb{C}, \quad T_p(f\otimes\lambda) = \lambda \operatorname{Sign}(g,A_p), \tag{5}$$

where $A_p = f^{-1}(p)$. We can prove that this is defined and independent of $\underline{p}$ and of the choice of $\underline{f}$ within its equivariant homotopy class. The proof is just the same as in the non-equivariant case (Milnor [31]). We choose an open simplex Δ^i of S^i (S^i has a fixed triangulation with respect to which $\underline{f}$ is simplicial) and, using Sard's theorem, a regular value $p \in \operatorname{Int}(\Delta^i)$ of $\underline{f}$. Then there is a homeomorphism (given explicitly in [31]) from $f^{-1}\Delta^i$ to $A_p\times\Delta^i$, commuting with the obvious maps to Δ^i, and from the definition this is a G-homeomorphism if $\underline{f}$ is G-equivariant. It follows that $T_p(f) = T_{p'}(f)$ for almost all $\underline{p}$ and all $\underline{p}'$ close to $\underline{p}$. Then a homotopy from $\underline{f}$ to a map $\underline{f}'$ gives a cobordism from $f^{-1}(p)$ to $f'^{-1}(p)$ for almost all $\underline{p}$, and this is a G-cobordism if the homotopy is G-equivariant. But the equivariant signature $\operatorname{Sign}(g,A)$ is an equivariant cobordism invariant of A (cf. Ossa [33]), so we deduce that $T_p(f) = T_p(f')$ for f,f' equivariantly homotopic. Since $\underline{f}$ is equivariantly homotopic to its composition with any simplicial automorphism of S^i, we can carry Δ^i onto any desired i-simplex of S^i without changing $T_p(f)$. Therefore $T_p(f) = T_{p'}(f')$ for f,f' equivariantly homotopic and almost all p,p' (now without requiring that $\underline{p}'$ be close to $\underline{p}$). This shows that the map T_p is well-defined, and that it is independent of $\underline{p}$ for almost all $\underline{p}$. We denote this common map by T. It is not hard to show that T is a homomorphism. Using the isomorphisms (4) and Poincaré duality in X/G (which is a rational homology manifold) we deduce the existence of $L\in H^*(X/G)$ with $T(f) = \langle \pi^*L \cup f^*\sigma, [X]\rangle$. Set $L(g,X) = \frac{1}{\deg \pi}\,\pi^*L \in H^*(X;\mathbb{C})^G$.

§5. The formula for $L(h',X/G)$

Now that we have defined $L(g,X)$ for rational homology G-manifolds, we find that the right-hand side of the formula for $L(X/G)$ proved in §3 in the differentiable case is also defined when X is a rational homology manifold, and we can ask whether it still gives the value of $L(X/G)$. This is the case, and we even have the following more general result, which was stated in somewhat other terms as Theorem 2 of §1:

Theorem 1: Let X be an oriented rational homology U-manifold, where U is a finite abelian group, and let G be a subgroup of U. Then the equivariant L-classes (in the sense of §4) of the induced action of U/G on the rational homology manifold X/G are given by the formula:

$$\frac{1}{\deg \pi}\ \pi^{*}L(\xi,X/G) = \frac{1}{|G|}\sum_{u\in\xi} L(u,X), \tag{1}$$

where $\xi \in U/G$ is a coset of G and π the projection $X \to X/G$.

Proof: Equality (1) is modelled after and proved using the corresponding equation for the equivariant signatures, namely

$$\frac{1}{\deg \pi}\ \mathrm{Sign}(\xi,A/G) = \frac{1}{|G|}\sum_{u\in\xi} \mathrm{Sign}(\mu,A), \tag{2}$$

where ξ is a coset of G in U and A is a U-manifold (or rational homology U-manifold; (2) is a purely homological statement). This is proved just as was the special case U=G (eq. (1) of §3) by applying to the positive and negative eigenspaces of the middle cohomology group of A the corresponding theorem for the traces $\mathrm{tr}(\xi|V^G)$ and $\mathrm{tr}(u|V)$ of the action of U on a vector space V and of the induced action of U/G on V^G.

To prove (1), we must show that the right-hand side is G-invariant (so that (1) defines an element of $H^*(X/G)$) and that the class $L(\xi,X/G)$ that it defines satisfies the basic equation (1) of §4. But we pointed out in §4 that equation (3) of that section holds also when X is a rational homology manifold; it follows that the right-hand side of (1) is G-invariant. (Here we do not need that U is abelian, which implies $h^*L(u,X) = L(u,X)$ for $h\in G$; it is sufficient if G is a normal subgroup of U, in which case the actions of G on the left and right in (1) change the summands on the right-hand side but leave the whole sum invariant. Theorem 1 is thus also true in this more general situation). Now we have to check that the class $L(\xi, X/G)$ defined by (1) satisfies (1) of §4. Choose a (U/G)-invariant map $\bar{f}$ from X/G to S^i and let $\bar{A}$ denote the inverse image of a sufficiently general point of S^i. Let $f = \bar{f} \circ \pi$ denote the corresponding U-invariant map $X \to S^i$ and A the inverse image of a point under S^i, so that $\bar{A} = A/G$. If we substitute $\frac{1}{\deg \pi} \pi_*[X]$ for $[X/G]$ in the definition of $L(\xi, X/G)$ we can rewrite this definition as

$$\frac{1}{\deg \pi} < \pi^* L(\xi, X/G) \cup \pi^* \bar{f}^* \sigma, [X] > \; = \; \mathrm{Sign}(\xi, \bar{A}), \qquad (3)$$

and if we then substitute for $\pi^* L(\xi, X/G)$ its value as given by (1), we obtain as the equation to be proved

$$\frac{1}{|G|} \sum_{u \in \xi} < L(u, X) \cup f^* \sigma, [X] > \; = \; \mathrm{Sign}(\xi, A/G). \qquad (4)$$

But the elements summed on the left-hand side of this are precisely the numbers Sign(u,A), again by eq. (1) of §4, and therefore the theorem has been reduced to the equality (2) given above.

§6. Application to a formula of Bott and some remarks on $L(g,X)$

To illustrate the behaviour of the equivariant L-class $L(g,X)$, we will calculate it in a simple case, namely for linear actions on $X = P_n\mathbb{C}$. Since this is a smooth action, we can calculate $L(g,X)$ by the Atiyah-Singer formula.

We write points of X as $(z_0:\ldots:z_n)$, where $(z_0,\ldots,z_n) \in S^{2n+1}$ is an $(n+1)$-tuple of complex numbers. Then the $(n+1)$-dimensional torus group T^{n+1} acts on X by coordinatewise multiplication, i.e. for

$$g = (\zeta_0,\ldots,\zeta_n) \in T^{n+1} = S^1 \times \ldots \times S^1 \tag{1}$$

we define the action on X by

$$g\circ(z_0:\ldots:z_n) = (\zeta_0 z_0:\ldots:\zeta_n z_n). \tag{2}$$

We must calculate the fixed-point set of g. Clearly this is

$$X^g = \{(z_0:\ldots:z_n) \mid \zeta_i z_i = \zeta z_i, \ i=0,\ldots,n \text{ for some } \zeta\}. \tag{3}$$

Since at least one of the numbers z_i is non-zero, the complex number ζ is uniquely determined by $z \in X$ and must belong to the finite subset $\{\zeta_0,\ldots,\zeta_n\}$ of S^1. Therefore we can write X^g as a finite disjoint union

$$X^g = \bigcup_{\zeta\in S^1} X(\zeta) \tag{4}$$

where

$$X(\zeta) = \{(z_0:\ldots:z_n) \in X \mid \zeta_i z_i = \zeta z_i, \ i=0,\ldots,n\}$$

$$= \{(z_0:\ldots:z_n) \mid z_i=0 \text{ for all } \underline{i} \text{ with } \zeta_i \neq \zeta\} \tag{5}$$

and

$$X(\zeta) = \emptyset, \qquad \text{if } \zeta \notin \{\zeta_0,\ldots,\zeta_n\}. \tag{6}$$

Therefore $X(\zeta)$ is isomorphic to a projective space $P_s(\mathbb{C})$, where $s+1$ is the number of indices $\underline{i}$ with $\zeta_i=\zeta$ (we set $s=-1$ in the case (6)). In particular $X(\zeta)$ is connected whenever it is nonempty, so that (4) is precisely the decomposition of X^g as the finite disjoint union of its connectedness components.

The value of $L(g,X)$ is now given by the Atiyah-Singer recipe as a sum over the components; we now calculate the contribution $L(g,X)_\zeta$ to $L(g,X)$ from a given component $X(\zeta)$. We let $\underline{s}$ denote the (complex) dimension of $X(\zeta)$, and for convenience renumber the coordinates so that

$$\zeta_0,\dots,\zeta_s = \zeta, \qquad \zeta_{s+1},\dots,\zeta_n \neq \zeta \; ; \tag{7}$$

then

$$X(\zeta) = \{(z_0:\dots:z_s:0:\dots:0)\} \approx P_s\mathbb{C} , \tag{8}$$

and we shall use this isomorphism to identify $X(\zeta)$ and $P_s\mathbb{C}$ without further comment. Let

$$x \in H^2(X), \qquad y \in H^2(X(\zeta)) \tag{9}$$

be the usual generators of the cohomology of complex projective space. Thus $y = c_1(H)$ where H is the Hopf bundle over $X(\zeta)$. Since the normal bundle of $X(\zeta)$ in X consists of $n-s$ copies of the Hopf bundle, its Chern class is $(1+y)^{n-s}$. Moreover, if we identify this normal bundle N^g with a tubular neighbourhood of $X(\zeta)$ in X, we see that the action of $\underline{g}$ is given by multiplication with $\zeta^{-1}\zeta_i$ on the ith copy of H. Indeed the fibre N^g_z of N^g at $z \in X$ is identified with $\mathbb{C}^{n-s}$ by the correspondence

$$(y_1,\dots,y_{n-s}) \leftrightarrow (z_0:\dots:z_s:y_1:\dots:y_{n-s})$$

and the action of $\underline{g}$ on N^g_z is therefore

$$\begin{aligned} g\circ(y_1,\dots,y_{n-s}) &\leftrightarrow g\circ(z_0:\dots:z_s:y_1:\dots:y_{n-s}) \\ &= (\zeta_0 z_0:\dots:\zeta_s z_s:\zeta_{s+1}y_1:\dots:\zeta_n y_{n-s}) \\ &= (\zeta z_0:\dots:\zeta z_s:\zeta_{s+1}y_1:\dots:\zeta_n y_{n-s}) \\ &= (z_0:\dots:z_s:\zeta^{-1}\zeta_{s+1}y_1:\dots:\zeta^{-1}\zeta_n y_{n-s}) \\ &\leftrightarrow (\zeta^{-1}\zeta_{s+1}\, y_1, \dots, \zeta^{-1}\zeta_n\, y_{n-s}). \end{aligned}$$

As was pointed out in §2, the Atiyah-Singer characteristic class $L_\pi(\xi)$ can be obtained from the same formula as that giving $L_\theta(\xi)$ if ξ is a complex bundle splitting up as a sum of complex line bundles. In our case N^g splits up into a sum of $n-s$ complex line bundles in an equivariant way, and each line bundle has characteristic class $\underline{y}$, while $X(\zeta)$ itself has total Chern class $(1+y)^{s+1}$ and therefore L-class $(y/\tanh y)^{s+1}$. Therefore the Atiyah-Singer formula (eq. (27) of §2) gives for the class $L'(g,X)_\zeta \in H^*(X(\zeta))$ the value

$$\begin{aligned} L'(g,X)_\zeta &= L(X(\zeta)) \prod_\theta L_\theta(N_\theta^g) \\ &= (y/\tanh y)^{s+1} \prod_{j=s+1}^{n} \frac{\zeta^{-1}\zeta_j e^{2y}+1}{\zeta^{-1}\zeta_j e^{2y}-1} \quad . \end{aligned} \tag{10}$$

Now it is clear that the Poincaré dual of $y^r \in H^{2r}(X(\zeta))$ is precisely the homology class represented by the submanifold $P_{s-r}\mathbb{C}$. If $\underline{j}$ denotes the inclusion $X(\zeta) \subset X$, we have $j_*[P_{s-r}\mathbb{C}] = [P_{s-r}\mathbb{C}]$, where the right-hand side denotes the homology class in X represented by the submanifold $P_{s-r}\mathbb{C} \subset P_n\mathbb{C}$. The Poincaré dual of this homology class is then in turn equal to $x^{n-s+r} \in H^{2n-2s+2r}(X)$. This shows that the Gysin homomorphism $j_!$ is given by

$$j_!(y^r) = x^{n-s+r}. \tag{11}$$

Thus to obtain $L(g,X)_\zeta$ from $L'(g,X)_\zeta$ we must replace $\underline{y}$ by $\underline{x}$ in formula (10) and then multiply the whole expression by x^{n-s}. This gives

$$\begin{aligned} L(g,X)_\zeta &= \left(\frac{x}{\tanh x}\right)^{s+1} \prod_{j=s+1}^{n} \left(x \, \frac{\zeta^{-1}\zeta_j e^{2x}+1}{\zeta^{-1}\zeta_j e^{2x}-1} \right) \\ &= \prod_{j=0}^{n} \left(x \, \frac{\zeta^{-1}\zeta_j e^{2x}+1}{\zeta^{-1}\zeta_j e^{2x}-1} \right) , \end{aligned} \tag{12}$$

where in the last line we have used the equality $\zeta_j=\zeta$ for $j=0,\dots,s$. Equation (12) is symmetric in the various coordinates, so the fact that we renumbered the coordinates at the beginning of the calculation does not matter, and (12) gives the desired contribution to $L(g,X)$ from the component $X(\zeta)$. If we now sum over the components, i.e. over all $\zeta \in S^1$ (that this is legitimate follows from the fact that (12) vanishes if $\zeta \notin \{\zeta_0,\dots,\zeta_n\}$, since then each of the n+1 factors is a power series in $\underline{x}$ beginning with a multiple of $\underline{x}$ rather than a constant term, and $x^{n+1} = 0$ in $H^*(X)$), we obtain:

<u>Theorem 1</u>: Let $g \in T^{n+1}$ act on $X = P_n\mathbb{C}$ by the action defined in (2), where $\zeta_0,\dots,\zeta_n \in S^1$ are complex numbers of norm 1. Then the equivariant

L-class $L(g,X)$ is given in terms of the Hopf class $x \in H^2(X)$ by

$$L(g,X) = \sum_{\zeta \in S^1} \prod_{j=0}^{n} \left(x \frac{\zeta^{-1}\zeta_j e^{2x} + 1}{\zeta^{-1}\zeta_j e^{2x} - 1} \right) . \tag{13}$$

The sum is in fact a finite one since the product appearing vanishes in $H^*(X)$ if $\zeta \notin \{\zeta_0,\dots,\zeta_n\}$.

Corollary: Let $\mu_a \subset S^1$ denote the cyclic subgroup of ath roots of unity, where a is a positive integer, and let

$$G = \mu_{a_0} \times\dots\times \mu_{a_n} = \{(\zeta_0,\dots,\zeta_n) \in T^{n+1} \mid \zeta_i^{a_i} = 1, i=0,\dots,n\} \tag{14}$$

be a finite subgroup of T^{n+1}, acting on $X = P_n\mathbb{C}$ as in the theorem, where $a_0,\dots,a_n$ are positive integers. Let d denote the greatest common divisor of the integers a_i. Let $p: X \to X/G$ be the projection map and all other notations as in Theorem 1. Then

$$p^*L(X/G) = \frac{1}{d} \sum_{0 \le \xi < \pi} \prod_{j=0}^{n} \frac{a_j x}{\tanh(a_j(x+i\xi))} . \tag{15}$$

Here the sum over all real numbers ξ between 0 and π is in fact finite, since the product vanishes unless $a_j\xi \equiv 0 \pmod{\pi}$ for at least one j.

Note: Formula (15) was originally proved by Bott (not yet published). It is quoted in Hirzebruch [16] (equation (1)) and a proof is given in [21].

Proof of corollary: By Theorem 1 of §1 we have

$$p^*L(X/G) = \frac{\deg p}{|G|} \sum_{g \in G} L(g,X), \tag{16}$$

and the factor $\frac{\deg p}{|G|}$ is simply the reciprocal of the order of the subgroup H of G of elements acting trivially on X. Clearly

$$H = \{(\zeta_0,\dots,\zeta_n) \in G \mid \zeta_0=\dots=\zeta_n\}$$

and it follows from the definition of G that $H = \mu_d$, where d is the greatest common divisor of $a_0,\dots,a_n$. We can evaluate the sum in (16) by using the trigonometric identity

$$\sum_{\lambda^a=1} \frac{\lambda z + 1}{\lambda z - 1} = a \frac{z^a + 1}{z^a - 1} \tag{17}$$

(which is proved by observing that the two sides have the same poles and the same residues). Thus

$$\sum_{g\in G} L(g,X) = \sum_{\substack{\zeta_0,\dots,\zeta_n \\ \zeta_i^{a_i}=1}} \sum_{\zeta\in S^1} \prod_{j=0}^{n} \left(x \frac{\zeta^{-1}\zeta_j e^{2x}+1}{\zeta^{-1}\zeta_j e^{2x}-1} \right)$$

$$= \sum_{\zeta\in S^1} \prod_{j=0}^{n} \left(x \cdot \sum_{\zeta_j^{a_j}=1} \frac{\zeta^{-1}\zeta_j e^{2x}+1}{\zeta^{-1}\zeta_j e^{2x}-1} \right)$$

$$= \sum_{\zeta\in S^1} \prod_{j=0}^{n} \left(a_j x \frac{\zeta^{-a_j} e^{2a_j x}+1}{\zeta^{-a_j} e^{2a_j x}-1} \right). \tag{18}$$

This sum is transformed into the one occurring in (15) by the substitution $\zeta = e^{-2i\xi}$, and the corollary is therefore proved.

We could of course write down an analogous formula for the equivariant L-class of the action of an element of T^{n+1}/G on the quotient space $P_n\mathbb{C}/G$ by using Theorem 2 instead of Theorem 1 of §1.

We will give a second application of Theorem 1 in §15. But the theorem is interesting for its own sake as well as for its applications, since it illustrates the behaviour of the equivariant L-class.

The first fact illustrated by (13) is that, even in the case of a differentiable action, $L(g,X) \in H^*(X;\mathbb{C})$ need not be a continuous function of g. In a way this is surprising, since $L(g,X)$ is defined using the equivariant signatures Sign(g,A), and the equivariant signature not only varies continuously but actually remains constant as one varies g (since it is determined by the action of g^* in cohomology). However, the submanifolds A on which $L(g,X)$ must be evaluated themselves depend on g (it is insufficient to know the value of $L(g,X)$ only on G-invariant submanifolds if G is an infinite group), and therefore one cannot conclude that $L(g,X)$ varies continuously. Indeed, we see that (13) has discontinuities at points $g\in G$ where two of the ζ_i's are equal for two reasons. First of all, the set $\{\zeta_0,\dots,\zeta_n\}$ over which the sum in (13) is to be taken becomes smaller when two of the ζ_i's coalesce; this discontinuity would not be there if one counted each $L(g,X)_\zeta$ with a multiplicity equal to the number of ζ_i's with $\zeta_i=\zeta$. But even if one were to do this, the right-hand side of (13) would still be discontinuous as a function of $(\zeta_0, \ldots, \zeta_n) \in T^{n+1}$, because the function

$$x \frac{\zeta^{-1}\zeta_j e^{2x} + 1}{\zeta^{-1}\zeta_j e^{2x} - 1} \tag{19}$$

is not continuous as a function of ζ at $\zeta = \zeta_j$. Indeed, although for each fixed ζ this is a continuous function (power series) in $\underline{x}$, its constant term (obtained by setting $x=0$) is zero if $\zeta = \zeta_j$ and one if $\zeta = \zeta_j$.

It is therefore certainly not the case that, for X a rational homology manifold, the function $L(g,X)$ defined in §4 for $g \in G' = \{\text{elements of finite order in } G\}$ is a continuous function from G' to $H^*(X;\mathbb{C})$. Therefore our first idea of extending the definition of $L(g,X)$ to all of G by using the denseness of G' cannot work. Nevertheless, it may be possible to define $L(g,X)$ by continuity. We state a

<u>Conjecture</u>: Let G be a compact abelian Lie group acting on a rational homology manifold X, G' the dense subgroup of elements of finite order, U the open subset $\{g\in G \mid X^g = X^G\}$, and $U' = U \cap G'$. Then the function $U' \to H^*(X;\mathbb{C})$ defined by the class $L(g,X)$ of §4 is continuous.

The point about this conjecture is that, one the one hand, it holds for differentiable actions and therefore has some hope of being true, and, on the other hand, it is strong enough to permit a consistent definition of $L(g,X)$ on all of G if it holds. To see the first statement, observe that the definition of $L(g,X)$ of §4 agrees with the previous definition §1(3) if X is a smooth G-manifold, and the Atiyah-Singer class $L'(g,X)$ clearly is a continuous function of $\underline{g}$ as long as the fixed-point set does not change. To see how the conjecture's truth would lead to a definition of $L(g,X)$ for all $g\in G$, we make the additional assumption that the action of G on X has only finitely many isotropy groups G_x (if X is an integral homology manifold, for instance, this is known to hold: cf. Borel [2], Ch. VI). Let $G_1, \ldots, G_k$ be the various isotropy groups occurring, and for each subset $I \subset \{1, \ldots, k\}$ define $G(I) = \cap_{i\in I} G_i$, $U(I) = \{g\in G \mid g\in G_i \Leftrightarrow i\in I\}$, and $X(I) = \{x\in X \mid G_x = G_i \text{ for some } i\in I\}$. Then G is the disjoint union of the 2^k sets $U(I)$, so it suffices to define $L(g,X)$ for $g\in U(I)$. But $G(I)$ is a closed subgroup of G, and $U(I) = \{g\in G(I) \mid X^g = X(I)\}$ is for $G(I)$ precisely the set U of the conjecture, so if the conjecture holds, the $L(g,X)$ of §4 is defined and continuous on a dense subset $U(I)'$ of $U(I)$ and can be extended to all of $U(I)$. As before, we note that assuming that G be abelian is harmless, since for arbitrary compact G we can define $L(g,X)$ by considering only the closed subgroup of G generated by *g*.

As well as revealing the discontinuity of the equivariant L-class $L(g,X)$, equation (13) is interesting because of a certain formal similarity with the main result of §5. To make this more apparent, we rename the spaces and groups involved. We now write X for S^{2n+1}, considered as the unit sphere in $\mathbb{C}^{n+1}$, let $U = T^{n+1} = S^1 \times \ldots \times S^1$ with the obvious action on X, and let G be the circle group S^1 embedded in U as the diagonal $(\zeta,\ldots,\zeta)$. Then the quotient $X/G = P_n\mathbb{C}$ is the space (1), and equation (13) can be restated in the form

$$L(\xi, X/G) = \sum_{u\in\xi} f(u) \qquad (\xi \in U/G = T^{n+1}/S^1), \tag{20}$$

where

$$f(u) = \prod_{j=0}^{n} \left(x \frac{\zeta_j e^{2x} + 1}{\zeta_j e^{2x} - 1} \right) \qquad (u = (\zeta_0,\ldots,\zeta_n) \in U). \tag{21}$$

Equation (20) is very similar to §5(2), which states that

$$\pi^* L(\xi, X/G) = \sum_{u\in\xi} L(u,X) \tag{22}$$

if G is finite and acts effectively on a rational homology manifold X. We might therefore conjecture that a formula like (20) holds in general, at least for G a compact abelian Lie group, if G acts freely (so that X/G is a manifold) or if only a finite number of elements of G have fixed points (e.g. for a semi-free S^1-action). The class $f(u)$ would then presumably satisfy $\pi^* f(u) = L(u,X)$, making (22) a consequence of (20), and would also have to be such that $f(u) = 0$ for all but finitely many <u>u</u> in each coset $\xi \in U/G$ (so that the sum (22) makes sense). I do not know if this is in general true. It is if G is finite abelian, since then $L(u,X)$ is invariant under G and we can simply take $f(u) = \pi^{*-1} L(u,X)$. It also is possible to define $f(u)$ if the action is differentiable, since then $(X/G)^\xi$ is the disjoint union of the sets X^u/G with $u\in\xi$ (and all but finitely many of these sets are empty) just as in (4), and so we can take for $f(u)$ the contribution in the Atiyah-Singer formula for $L(\xi, X/G)$ coming from the component X^u/G of the fixed-point set $(X/G)^\xi$.

CHAPTER II: L-CLASSES OF SYMMETRIC PRODUCTS

In Chapter I we obtained a formula for the calculation of the L-class of the quotient of a differentiable manifold by a smooth, orientation-preserving action of a finite group. In this chapter, we apply this to the quotient X^n/S_n of the nth Cartesian product of X with itself by the symmetric group on n letters, i.e. to the nth symmetric product X(n) of X. In order that the action of S_n on X^n be orientation-preserving, it is necessary that X have even dimension 2s.

We give in §7 a description of the rational cohomology of X(n), as well as the various notation necessary for a statement of the formula for L(X(n)), which is then given in the following section. Since the complete formula obtained is rather obscure, we also make several comments explaining the meaning of the various terms and of the formula as a whole. The proof occupies the following four sections. Section 13 contains an explicit evaluation of the formula in two special cases-- X a sphere of even dimension, and X a Riemann surface. In the latter case, X(n) is a complex manifold whose Chern class (and hence also L-class) was already known (Macdonald [26]); thus we are able to check our general formula.

In §14 we consider the more general situation where a finite group G acts on X. Then there is an induced action on X(n) (diagonal action), and we can apply the result of §5 to calculate L(g,X(n)) for this action. Since the calculation is similar to that in the non-equivariant case, we give the proof more briefly, in a single section. The formula obtained is even more complicated than the non-equivariant one, but again we find that the dependence of L(g,X(n)) on n is very simple--if the order of g is odd. In the last section of the chapter we compute explicitly the form which the equation for L(g,X(n)) takes if X is an even-dimensional sphere. If X is S^2, then $X(n) = P_n\mathbb{C}$ and we can check the result with the computation of $L(g,P_n\mathbb{C})$ made previously in connection with the theorem of Bott.

The most interesting fact which emerges is the very simple dependence of L(X(n)) on the number of factors n in the symmetric power. Namely, if we consider the inclusion j of X(n) in X(n+1) induced by choosing a base-point in X, and then form $Q_n = j^*L(X(n+1))\cdot L(X(n))^{-1}$ (this is possible since L(X(n)) is invertible; Q_n would be the L-class of the normal bundle if $X(n)$ were a smooth manifold smoothly embedded in $X(n+1)$), it turns out that Q_n is independent of $\underline{n}$ in the sense that $j^*Q_{n+1} = Q_n$. Thus Q_n is the restriction to $X(n)$ of a class Q in the cohomology of $X(\infty)$. Moreover, Q turns out to be in a certain sense independent of X. There is namely a canonically defined element $\eta \in H^{2s}(X(\infty))$ arising from the orientation of X, and the class Q is equal to $Q_s(\eta)$ with Q_s a power series in one variable whose coefficients depend only on $s = (\dim X)/2$. For example, $Q_1(t) = t/\tanh t$; this corresponds to the case where X is a Riemann surface, in which case $X(n)$ is a smooth manifold and Q_n really can be interpreted as the L-class of the normal bundle of $X(n) \subset X(n+1)$. For $\underline{s}$ larger than 1, however, the power series $Q_s(t)$ does not split formally as a finite product $\Pi_{j=1}^{s}(x_j/\tanh x_j)$, and we deduce that the inclusion of $X(n)$ in $X(n+1)$ does not have a normal bundle in the sense of Thom.

In the equivariant case, if $\underline{g}$ acts on X and has finite order $\underline{p}$, we get a similar result if $\underline{p}$ is odd, namely the restriction to $H^*(X(n))$ of $L(g,X(n+p))$ equals the product of $L(g,X(n))$ with a factor Q which is independent of $\underline{n}$, of X, and of the action of $\underline{g}$ on X, but depends only on $\underline{s}$ and $\underline{p}$ (it is again a power series in η, this time with coefficients depending on $\underline{p}$ as well as on $\underline{s}$; for instance it equals $\eta/\tanh p\eta$ if $s=1$). This result has two features of interest: first, that there is a simple relation between $L(X(n))$ and $L(X(n+p))$, so that there is a kind of periodicity with period $\underline{p}$ in the equivariant structure of the symmetric products of X, and secondly, that there is a different behaviour for elements of odd and even order (if $\underline{p}$ is even there is no periodicity).

Since the calculations are long and perhaps unconvincing, I have tried to give as many computations as possible for which a known result was available for comparison. Unfortunately, this was only possible for X two-dimensional, since the power series $Q_s(\eta)$ for $s>1$ has not previously occurred in this connection. As already mentioned, it was possible in two cases: Macdonald's work on symmetric powers of Riemann surfaces for the non-equivariant case, and the formula of §6 (here $X=S^2$) for the equivariant case. In both cases, the previous result had been obtained by quite different methods (Macdonald does not use the index theorem, and in §6 we did not use the fact that $P_n\mathbb{C}$ is a symmetric product) and in a quite different form requiring a long computation to be shown equal to our result. Thus these verifications lend considerable credibility to the theorems of this chapter.

§7. The rational cohomology of $X(n)$

If X is a compact, connected topological space and n a positive integer, the nth symmetric product of X is the space

$$X(n) = X^n/S_n \tag{1}$$

with the quotient topology, where S_n is the symmetric group on n letters acting on the nth Cartesian product X^n of X with itself by permutation of the factors. Thus $X(n)$ is the set of all unordered n-tuples of points of X, with the obvious topology. If $n=0$, we define $X(n)$ to be a point. The nth symmetric product of X is also sometimes called its nth symmetric power, and is variously denoted $X(n)$, $X[n]$, and SP^nX.

If we choose a base-point $x_0 \in X$, then there is a natural inclusion

$$j: X(n) \subset X(n+1) \tag{2}$$

sending an unordered n-tuple $\{x_1, \ldots, x_n\} \subset X$ to its union $\{x_0, x_1, \ldots, x_n\}$ with $\{x_0\}$. However, there is no natural projection map from $X(n+1)$ to $X(n)$, since there is no natural way to choose n elements from a set of $n+1$ elements. We will also use the letter j to denote compositions of the map (2) with itself, i.e. any inclusion $X(n) \subset X(m)$ with $m > n$. Let

$$X(\infty) = \varinjlim X(n) \tag{3}$$

be the limit of the direct system defined by the maps j, and use j to denote the inclusion of $X(n)$ in $X(\infty)$ also.

The purpose of this section is to describe the (additive) rational cohomology of $X(n)$. This is very easy, using $H^*(X/G;\mathbb{Q}) \cong H^*(X;\mathbb{Q})^G$ and elementary properties of cohomology (as found in Spanier), and this section can be skipped except for the equations (8), (9), (17) giving the notations used for the basis of $H^*(X(n);\mathbb{Q})$ and for one easy proposition (eq. (24)).

All cohomology in this section is to be understood with coefficients in $\mathbb{Q}$ (or $\mathbb{R}$ or $\mathbb{C}$; we only need to have a field of characteristic zero). This results in two simplifications. First, as mentioned above, the map $\pi^*: H^*(X/G) \to H^*(X)$ is then an isomorphism onto $H^*(X)^G$ (cf. §1); this isomorphism will be used without explicit mention to identify $H^*(X/G)$ with $H^*(X)^G \subset H^*(X)$ for any G-space X with G a finite group.

Secondly, there is a natural isomorphism of $H^*(X\times Y)$ with $H^*(X) \otimes H^*(Y)$. Therefore

$$H^*(X^n) = H^*(X) \otimes \ldots \otimes H^*(X) \tag{4}$$

has as a basis $\{f_{i_1}\times\ldots\times f_{i_n} \mid i_1,\ldots,i_n \in I\}$, where $\{f_i \mid i \in I\}$ is an additive basis for the finite-dimensional vector space $H^*(X)$ over $\mathbb{Q}$. Recall that the map in cohomology induced by the interchange mapping

$$T: X\times Y \to Y\times X \tag{5}$$

sends $v\times u \in H^{j+i}(Y\times X)$ (where $u \in H^i(X)$, $v \in H^j(Y)$) to $(-1)^{ij} u\times v \in H^{i+j}(X\times Y)$. In particular, with $X=Y$ we see that the map induced in cohomology by an interchange of factors is not simply the corresponding interchange, but contains a further factor -1 if two terms of odd degree are transposed. It follows that, for $\sigma \in S_n$, the effect of σ^* on $H^*(X^n)$ (where σ acts on X^n by $\sigma(x_1,\ldots,x_n) \to (x_{\sigma(1)},\ldots,x_{\sigma(n)})$) is

$$\sigma^*(u_1\times\ldots\times u_n) = (-1)^\nu\, u_{\sigma^{-1}(1)}\times\ldots\times u_{\sigma^{-1}(n)}, \tag{6}$$

where $u_1,\ldots,u_n$ in $H^*(X)$ are homogeneous elements and ν is the number of transpositions $i\leftrightarrow j$ of the permutation σ for which u_i and u_j have odd degree (this number is well-defined, i.e. independent of the decomposition of σ as a product of transpositions, modulo 2). Write $\pi_j: X^n \to X$ for the projection onto the jth factor; then we can reformulate (6) as

$$\begin{aligned}\sigma^*(u_1\times\ldots\times u_n) &= (-1)^\nu\, \pi_1^*(u_{\sigma^{-1}(1)})\cup\ldots\cup\pi_n^*(u_{\sigma^{-1}(n)}) \\ &= \pi_{\sigma(1)}^* u_1 \cup\ldots\cup \pi_{\sigma(n)}^* u_n,\end{aligned} \tag{7}$$

where in the last line we have used graded commutativity. (One can see this more easily by noting that $\pi_j\circ\sigma = \pi_{\sigma(j)}$, so that $\sigma^*(\pi_j^* u_j) = \pi_{\sigma(j)}^* u_j$.) Therefore the symmetrization of $u_1\times\ldots\times u_n$ is

$$\begin{aligned}\langle u_1,\ldots,u_n\rangle &= \sum_{\sigma\in S_n} \sigma^*(u_1\times\ldots\times u_n) = \sum_{\sigma\in S_n} \pi_{\sigma(1)}^* u_1 \ldots \pi_{\sigma(n)}^* u_n \\ &\in H^*(X^n)^{S_n} = H^*(X(n)).\end{aligned} \tag{8}$$

Here and in the future, we omit the symbol $\cup$ and denote cup products simply by juxtaposition.

We choose a basis $f_0,\ldots,f_b$ of $H^*(X)$ with $f_b=1$ and each f_i

homogeneous, say of degree d_i. Then $H^*(X(n))$ is spanned by the elements $<f_{j_1},\dots,f_{j_n}>$. Since this element is independent (up to sign) of the order of the indices j_i, we need only specify the number n_j of times a given index j $(j=0,1,\dots,b)$ appears in the set $\{j_1,\dots,j_n\}$. Thus, let

$$<n_0(f_0)\dots n_b(f_b)> = <\underbrace{f_0,\dots,f_0}_{n_0},\underbrace{f_1,\dots,f_1}_{n_1},\dots,\underbrace{f_b,\dots,f_b}_{n_b}> \in H^*(X(n)), \tag{9}$$

where $n_0+\dots n_b=n$. There is, however, one more restriction if we wish to have a basis: namely, n_i must be ≤ 1 if $d_i=\deg f_i$ is odd. Indeed, from (8) we see immediately that if, for example, $u_1=u_2$ and is of odd degree, then $<u_1,\dots,u_n>$ is zero. Thus we have

<u>Proposition 1</u>: Let $f_0,\dots,f_b$ be a homogeneous basis for $H^*(X;\mathbb{Q})$. Then a basis for $H^*(X(n);\mathbb{Q})$ is given by the set of elements (9) with $n_0+\dots+n_b = n$ and $n_i \leq 1$ for all i with $\deg f_i$ odd.

<u>Corollary</u>: The Poincaré polynomial

$$P(X(n)) = \sum_{r=0}^{\infty} x^r \dim_{\mathbb{Q}} H^r(X(n);\mathbb{Q}) \tag{10}$$

of $X(n)$ is given in terms of the Betti numbers

$$\beta_d = \dim_{\mathbb{Q}} H^d(X;\mathbb{Q}) \tag{11}$$

of X by the formula

$$\sum_{n=0}^{\infty} t^n P(X(n)) = \prod_{\substack{d\geq 0\\ d \text{ even}}} \left(\frac{1}{1-tx^d}\right)^{\beta_d} \prod_{\substack{d\geq 0\\ d \text{ odd}}} \left(1+tx^d\right)^{\beta_d}. \tag{12}$$

In particular $(x=-1)$, the Euler characteristics of $X(n)$ and X are related by

$$\sum_{n=0}^{\infty} t^n e(X(n)) = \left(\frac{1}{1-t}\right)^{e(X)}. \tag{13}$$

<u>Note</u>: Formulas (12) and (13) are due to Macdonald [24].

<u>Proof of corollary</u>: By the proposition, we have

$$\dim_{\mathbb{Q}} H^r(X(n)) = \#\{n_0,\dots,n_b \geq 0 \mid n_0+\dots+n_b = n,\ n_i \leq 1 \text{ for } d_i \text{ odd},\ n_0d_0+\dots+n_bd_b = r\},$$

from which we find:

$$\sum_{n=0}^{\infty}\sum_{r=0}^{\infty} t^n x^r \dim H^r(X(n)) = \sum_{\substack{n_0,\dots,n_b\geq 0\\ n_i\leq 1 \text{ for } d_i \text{ odd}}} t^{n_0+\dots+n_b} x^{n_0d_0+\dots+n_bd_b}$$

$$= \prod_{\substack{j=0\\ d_j \text{ even}}}^{b}\left(\sum_{n=0}^{\infty} t^n x^{nd_j}\right)\prod_{\substack{j=0\\ d_j \text{ odd}}}^{b}\left(\sum_{n=0}^{1} t^n x^{nd_j}\right)$$

$$= \prod_{\substack{j=0\\ d_j \text{ even}}}^{b}\left(\frac{1}{1-tx^{d_j}}\right)\prod_{\substack{j=0\\ d_j \text{ odd}}}^{b}\left(1+tx^{d_j}\right)$$

$$= \prod_{\substack{d=0\\ d \text{ even}}}^{\infty}(1-tx^d)^{-\beta_d}\prod_{\substack{d=0\\ d \text{ odd}}}^{\infty}(1+tx^d)^{\beta_d} . \qquad (14)$$

We now want to compare the cohomology of $X(n)$ with that of $X(n+1)$, using the inclusion (2). The inclusion j' of X^n in X^{n+1} as $X^n \times x_0$ induces the map

$$j'^*(u_1\times\dots\times u_{n+1}) = \begin{cases} u_1\times\dots\times u_n, & \text{if } u_{n+1} = 1\\ 0, & \text{if } u_{n+1} \neq 1\end{cases} \qquad (15)$$

in cohomology; therefore the map j^* from $H^*(X(n+1)) = H^*(X^{n+1})^{S_{n+1}}$ to $H^*(X(n)) = H^*(X^n)^{S_n}$, which is simply the restriction of j'^* to the S_{n+1}-invariant subspace of $H^*(X^{n+1})$, operates by

$$j^*\langle u_1,\dots,u_{n+1}\rangle = \sum_{\substack{\sigma\in S_{n+1}\\ u_{\sigma^{-1}(n+1)}=1}} \pi^*_{\sigma(1)}(u_1)\dots\widehat{\pi^*_{n+1}(u_{\sigma^{-1}(n+1)})}\dots\pi^*_{\sigma(n+1)}(u_{n+1})$$

$$= \sum_{\substack{j=1\\ u_j=1}}^{n+1}\ \sum_{\substack{\sigma\in S_{n+1}\\ \sigma(j)=n+1}} \pi^*_{\sigma(1)}\, u_1 \dots \widehat{\pi^*_{n+1}\, u_j} \dots \pi^*_{\sigma(n+1)}\, u_{n+1}$$

$$= \sum_{\substack{j=1\\ u_j=1}}^{n+1} \langle u_1,\dots,\widehat{u_j},\dots,u_{n+1}\rangle ,$$

where as usual the ^ indicates the omission of a factor. Therefore

$$j^* \langle n_0(f_0)\ldots n_b(f_b)\rangle \quad \begin{cases} n_b \langle n_0(f_0)\ldots n_{b-1}(f_{b-1})\overline{n_b-1}(f_b)\rangle & \text{if } n_b>0, \\ 0 & \text{if } n_b=0, \end{cases}$$

or

$$j^*\left(\frac{1}{n_b!}\langle n_0(f_0)\ldots n_b(f_b)\rangle\right) = \frac{1}{(n_b-1)!}\langle n_0(f_0)\ldots n_{b-1}(f_{b-1})\overline{n_b-1}(f_b)\rangle. \tag{16}$$

Thus if we renormalize the element (9) by dividing by $n_b!$, we obtain an element which is stable under j^*. We therefore define

$$[n_0(f_0)\ldots n_{b-1}(f_{b-1})]_n = \begin{cases} 0, & \text{if } n < n_0+\ldots+n_{b-1}, \\ (n_b!)^{-1}\langle n_0(f_0)\ldots n_{b-1}(f_{b-1})n_b(f_b)\rangle, & \text{if } n \geq n_0+\ldots+n_{b-1}, \end{cases}$$

$$\in H^*(X(n)), \tag{17}$$

where in the case $n \geq n_0+\ldots+n_{b-1}$ we have defined n_b as $n-n_0-\ldots-n_{b-1}$. Then (16) states

$$j^*[n_0(f_0)\ldots n_{b-1}(f_{b-1})]_{n+1} = [n_0(f_0)\ldots n_{b-1}(f_{b-1})]_n . \tag{18}$$

In other words, the sequence $[n_0(f_0)\ldots n_{b-1}(f_{b-1})]_n$ $(n=1,2,\ldots)$ defines an element

$$[n_0(f_0)\ldots n_{b-1}(f_{b-1})] \in H^*(X(\infty)) = \varprojlim_n H^*(X(n)) \tag{19}$$

such that

$$j^*[n_0(f_0)\ldots n_{b-1}(f_{b-1})] = [n_0(f_0)\ldots n_{b-1}(f_{b-1})]_n \tag{20}$$

for the inclusion $j\colon X(n) \subset X(\infty)$. Because of relation (20), we will omit the subscript $\underline{n}$ in future, leaving the question whether the stable element or its restriction to some $X(n)$ is meant to be decided by the context. Notice that, since $f_0,\ldots,f_{b-1}$ have positive degrees, there are only finitely many elements (19) of given degree, so that each of the groups $H^j(X(\infty))$ is finite-dimensional over $\mathbb{Q}$ (though infinitely many of them are non-zero). In particular $X(\infty)$ has a well-defined Poincaré power series, which from eq. (12) above is clearly

$$P(X(\infty)) = \prod_{d>0,\ \text{even}} (1-x^d)^{-\beta_d} \prod_{d>0,\ \text{odd}} (1+x^d)^{\beta_d} . \tag{21}$$

We have looked especially closely at the rôle of $f_b=1$ in expression (9) above. However, there are other f_i's in (9) which act in a special way, namely those of the highest degree occurring (or, more generally, any f_i such that $f_i f_j=0$ for all f_j of positive degree). Assume that f_0 is such an element, and that its degree $\underline{d}$ is $\underline{\text{even}}$ (this is the only case we will need later). Define

$$\eta_n = [1(f_0)]_n = \sum_{i=1}^{n} \pi_i^* f_0 \in H^d(X(n)). \tag{22}$$

Then

$$\eta_n < u_1,\ldots,u_n > = \sum_{i=1}^{n} \sum_{\sigma\in S_n} (\pi_i^* f_0)(\pi_{\sigma(1)}^* u_1)\ldots(\pi_{\sigma(n)}^* u_n).$$

In the expression on the right, we interchange the summations and replace i by $j=\sigma(i)$, obtaining (since f_0 is of even degree and commutes with everything)

$$\sum_{\sigma\in S_n} \sum_{j=1}^{n} (\pi_{\sigma(1)}^* u_1)\ldots(\pi_{\sigma(j)}^*(u_j f_0))\ldots(\pi_{\sigma(n)}^* u_n).$$

But $u_j f_0$ is zero unless $u_j=1$, by our assumption on f_0, so this is

$$\sum_{\substack{j=1\\ u_j=1}}^{n} \left[\sum_{\sigma\in S_n} (\pi_{\sigma(1)}^* u_1)\ldots(\pi_{\sigma(j)}^* f_0)\ldots(\pi_{\sigma(n)}^* u_n) \right]$$

$$= \sum_{\substack{j=1\\ u_j=1}}^{n} <u_1,\ldots,u_{j-1},f_0,u_{j+1},\ldots,u_n>.$$

Therefore (we use an overline rather than brackets to avoid ambiguity)

$$\eta_n < n_0(f_0)\ldots n_{b-1}(f_{b-1}) n_b(f_b) > = n_b < \overline{n_0+1}(f_0)\ldots n_{b-1}(f_{b-1})\overline{n_b-1}(f_b)>,$$

or, finally,

$$\eta_n\, [n_0(f_0)\ldots n_{b-1}(f_{b-1})]_n = [\overline{n_0+1}(f_0) n_1(f_1)\ldots n_{b-1}(f_{b-1})]_n. \tag{23}$$

Clearly the sequence η_n $(n=1,2,\ldots)$ defines an element $\eta \in H^d(X(\infty))$ and (23) is a stable relation. We have thus proved

<u>Proposition 2</u>: Let $\deg f_0 = d$ be even and $d \geq d_i$ for all i. Then

$$[n_0(f_0) n_1(f_1)\ldots n_{b-1}(f_{b-1})] = \eta^{n_0}\, [n_1(f_1)\ldots n_{b-1}(f_{b-1})] \tag{24}$$

in $H^*(X(\infty))$, where $\eta = [1(f_0)] \in H^d(X(\infty))$.

§8. Statement and discussion of the formula for $L(X(n))$

We now assume that the space X whose symmetric powers are being studied is a closed, connected, oriented differentiable manifold. Then the action of the symmetric group S_n on X^n is smooth; the condition that X^n/S_n be a rational homology manifold is that this action is also orientation-preserving. Since S_n is generated by transpositions (for $n > 1$), this is the case exactly when the interchange map $T: X\times X \to X\times X$ is orientation-preserving. Because of the graded commutativity, this holds if and only if the dimension of X is an even number $2s$.

We let

$$z \in H^{2s}(X) \tag{1}$$

be the Poincaré dual of $[\text{pt.}] \in H_0(X)$. We then choose a basis for $H^*(X)$ as in §7, i.e. $f_0, \dots, f_b$ with each f_i homogeneous and with $f_0 = z$ and $f_b = 1$. We also introduce a second basis

$$e_0, \dots, e_b \in H^*(X), \quad e_0 = 1, \quad e_b = z, \tag{2}$$

which is dual to $\{f_i\}$ under the intersection form, that is,

$$< e_i f_j, [X] > \; = \; \delta_{ij} \quad (i, j = 0, 1, \dots, b). \tag{3}$$

Thus e_i is homogeneous of degree $2s-d_i$, where $d_i = \deg f_i$.

The result of the last section was that a basis for $H^*(X(\infty))$ is given by the elements

$$[n_0(f_0)\dots n_{b-1}(f_{b-1})] \; = \; \eta^{n_0}\, [n_1(f_1)\dots n_{b-1}(f_{b-1})], \tag{4}$$

where the square brackets have the meaning given in §7 and where

$$\eta = [1(f_0)] \in H^{2s}(X(\infty)) \tag{5}$$

is the element whose restriction to $X(n)$ is

$$\eta_n = j^*\eta = \pi_1^* z + \dots + \pi_n^* z \in H^{2s}(X(n)). \tag{6}$$

The map $j^*: H^*(X(\infty)) \to H^*(X(n))$ is surjective, and a basis for $H^*(X(n))$ is given by the images under j^* of the elements (4) for which $n_0 + \dots + n_{b-1} \leq n$.

We now state the main theorem of this chapter.

Theorem 1: Let X be a connected, closed, oriented differentiable manifold of dimension 2s. Let η be the element defined in (5) and (6). Let j denote the inclusion of X(n) in X(n+1). Then

$$j^*L(X(n+1)) = Q_s(\eta_n)\cdot L(X(n)), \tag{7}$$

where $Q_s(-)$ is the power series in one variable defined below, and depends on s but not on the space X or the number of factors n. An equivalent statement is that there exists an element G of $H^{**}(X(\infty))$ such that (if we use j also to denote the inclusion $X(n)\subset X(\infty)$)

$$L(X(n)) = j^*[Q_s(\eta)^{n+1}\cdot G]\ . \tag{8}$$

The power series Q_s is defined as the unique even power series such that

$$\text{coefficient of } t^{2k} \text{ in } Q_s(t)^{2k+1} = (2k+1)^{-s+1} \qquad (k\geq 0). \tag{9}$$

It can also be defined as

$$Q_s(t) = \frac{t}{f_s(t)}\ , \tag{10}$$

where

$$f_s(t) = g_s^{-1}(t), \qquad g_s(t) = t + \frac{t^3}{3^s} + \frac{t^5}{5^s} + \frac{t^7}{7^s} + \dots\ . \tag{11}$$

The first few terms of these power series are given in the table at the end of this section.

The class $G\in H^{**}(X(\infty))$ is defined as follows: Let B be the graded power series ring over $\mathbb{Q}$ in variables $t_0,\dots,t_{b-1}$ with t_i of degree d_i, i.e. B is the quotient of $\mathbb{Q}[[t_0,\dots,t_{b-1}]]$ by the relations

$$t_i t_j = (-1)^{d_i d_j}\, t_j t_i \qquad (i,j = 0,1,\dots,b-1). \tag{12}$$

Let A be the graded tensor product of $H^{**}(X)$ with B. We use e to denote $e\otimes 1$ (for $e\in H^{**}(X)$) and t_i to denote $1\otimes t_i$ $(i=0,\dots,b-1)$; then

$$e\,t_i = (-1)^{dd_i}\, t_i e \qquad (\,e\in H^d(X)\,) \tag{13}$$

in $\mathbf{A}$. We define $\alpha\in\mathbf{A}$ by

$$\alpha = e_0 t_0 + \dots + e_{b-1} t_{b-1}\ . \tag{14}$$

Since B is a power series ring rather than a polynomial ring and we tensored with $H^{**}(X)$ instead of $H^*(X)$ in forming A, it is legitimate to form power series (with rational coefficients) of elements in A; thus $g_r(\alpha) \in A$ is well-defined. We write $L_c(X)$ for the component of degree 2c of the L-class of X (<u>this deviates from the usual notation</u> where L_c denotes a certain polynomial in Pontrjagin classes of total degree 4c), or for the corresponding element $L_c(X)\otimes 1$ of A. There is a linear map $\langle -, [X]\rangle$ from the graded ring A to the graded ring B, defined on the generators $e\, t_{i_1}\ldots t_{i_k}$ of A by

$$\langle e\, t_{i_1}\ldots t_{i_k}, [X]\rangle \quad = \quad \langle e,[X]\rangle \; t_{i_1}\ldots t_{i_k} \in B. \tag{15}$$

Set

$$H \quad = \quad \Sigma_{c=0}^{s} \langle g_{s+1-c}(\alpha) L_c(X), [X]\rangle \quad \in \quad B \tag{16}$$

and define numbers $c_{n_0\ldots n_{b-1}} \in \mathbb{Q}$ as the coefficients in the expansion

$$\frac{e^H}{(1-t_0^2)^{e(X)/2}} \quad = \quad \sum c_{n_0\ldots n_{b-1}} \, t_0^{n_0}\ldots t_{b-1}^{n_{b-1}} \quad \in B, \tag{17}$$

where $e(X)$ is the Euler characteristic, e^H denotes exponentiation, and the sum is over all integers $n_0,\ldots,n_{b-1} \geq 0$. Then

$$G \;=\; f_s'(\eta) \sum c_{n_0\ldots n_{b-1}} \, Q_s(\eta)^{-n_0-\ldots-n_{b-1}} \, [n_0(f_0)\ldots n_{b-1}(f_{b-1})], \tag{18}$$

the summation being as in (17), where $Q_s(-)$ and $f_s(-)$ are the power series defined above and $f_s'(-)$ is the (formal) derivative of $f_s(-)$.

Before discussing the interpretation or significance of Theorem 1, we should perhaps make some remarks elucidating the formalism, since the complete statement is somewhat involved and perhaps confusing.

In the first place, the essential and most interesting assertion of the theorem is the relatively simple statement (7), either as a pure existence theorem (namely the existence of a power series Q_s independent of <u>n</u> for which (7) holds) or, combined with the formula for Q_s given in eq. (9) or in eqs. (10)-(11), as a precise statement of the relationship between the values of $L(X(n))$ for two successive values of <u>n</u>. The interest of this relation lies in a certain formal resemblance to a similar relation involving the L-class of a normal bundle; this will be discussed below. As stated in the theorem, equation (7) is equivalent to the existence of a cohomology class G in $X(\infty)$ such that (8) holds for every n. To see this, we notice that (8) trivially implies (7) since the inclusion of $X(n)$ in $X(\infty)$ is the composition $X(n) \subset X(n+1) \subset X(\infty)$. Conversely, if (7) holds for every n, we define a class G_n in $H^*(X(n))$ as $Q_s(\eta_n)^{-n-1} L(X(n))$ and deduce from (7) that $j^* G_{n+1} = G_n$, which means that the sequence $\{G_n\}_{n=1}^{\infty}$ defines an element G of the inverse limit $\varprojlim H^*(X(n)) = H^*(X(\infty))$ with $j^* G = G_n$ for all n. Thus we can break the theorem up into the main statement, that $L(X(n))$ behaves exponentially as a function of n (i.e. is of the form $Q^{n+1} G$ with Q, G independent of n), the statement that Q is given by the expressions (9)-(11), and the statement that G is given by the procedure stated in the remainder of the theorem.

The definition of the power series $Q_s(t) = 1 + t^2/3^s + \dots$ (the first few coefficients of Q_s and various related power series are given in the table at the end of this section) requires little comment. The equivalence of the definition (9) and the definition given in eqs. (10) and (11) is of course an exact parallel to the characterisation of the function

$$Q_1(t) = \frac{t}{\tanh t} \tag{19}$$

given by Hirzebruch ([12], Lemma 1.5.1), and the proof is exactly the same. The functions $Q_0(t)$ and $Q_1(t)$ can be written in closed form ($Q_0(t)$ is equal to $[1 + \sqrt{1+4t^2}]/2$) but the series $Q_s(t)$ for $s \geq 2$ is not an elementary function. Note that, since $Q_s(t)$ is an even power series and η has degree 2s, equation (7) implies that $L(X(n))$ and $L(X(n+1))$ agree up to degree 4s (or more precisely, $j^* L_i(X(n+1)) = L_i(X(n))$ for $i < 2s$), so the value of $L(X(n))$ is essentially independent of n in degrees smaller than $2 \cdot \dim X$.

Even less need be said to explain the definition of G, though this is more complicated. We only remark that, since $g_r(t)$ is a power series with no constant term, H considered as a power series in the t_i's also has no constant term, so it is legitimate to form the power series e^H in equation (17).

It would be very pleasing to have a simple proof of equation (7), or equivalently of (8) without an explicit evaluation of the power series G. One way to do this might be to assume given the class $L(X(n+1))$ and show directly that $Q_s(\eta_n)^{-1}\ j^*L(X(n+1))$ satisfies Milnor's definition for the L-class of $X(n)$.

However, despite the rather cumbersome formula for G, it is possible in certain situations to use the entire formula (8) to obtain completely explicit formulas for $L(X(n))$. This is the case when X has especially simple cohomology and L-class, for example if it is a sphere (of even dimension) or a Riemann surface (for these two cases the calculation will be given in §13). Moreover, it is possible to write G in a shorter and more natural way--in particular, in such a way that it does not involve the choice of a basis for $H^*(X)$--and indeed it is this basis-free version which will emerge from the proof in §§9-12. We preferred the more unwieldy form given here because the basis-free formulation requires the introduction of yet more abstract notation which would have made the statement of the theorem even more obscure, and because one is in any case obliged to expand in terms of a basis for concrete applications such as those mentioned above (indeed, even the power series $g_r(\alpha)$ and e^H used to shorten the statement of the theorem given here must be expanded in the course of an explicit calculation).

We can use the evaluation of G to obtain complete results for $L(X(n))$ not only when X is a simple manifold, but also for arbitrary X and small values of n. This is slightly easier using the basis-free form of Theorem 1 but can also be done with the form given here. For n=1 we can easily check that Theorem 1 really does give $L(X)$ as the value of $L(X(1))$. For n=2 we obtain (with $\pi: X\times X \to X(2)$ the projection)

$$\pi^*L(X(2)) \quad = \quad L(X)\times L(X) \ + \ e(X)\ z\times z. \tag{20}$$

In particular, the signature of the symmetric square of X equals $[(\text{Sign}\ X)^2 + e(X)]/2$, a special case of a formula of Hirzebruch for $\text{Sign}(X(n))$ which will be proved in §9 (and deduced from the formula for $L(X(n))$ as a check on the latter in §13). For $n\geq 3$ the formula for $\pi^*L(X(n))$ is rather more complicated; it is a certain polynomial in the lifts to $H^*(X^n)$ of the elements $z\in H^{2s}(X)$, $L(X)\in H^*(X)$, and $a\in H^{2s}(X\times X)$, this last being the restriction to $X\times X$ of the Thom class U in $H^{2s}(X\times X, X\times X-\Delta)$ (where Δ is the diagonal in $X\times X$).

The fact that $L(X(n))$ is given--when X is a differentiable manifold--by a certain polynomial in the classes $\underline{z}$, $L_0(X)$, and $\underline{a}$ and their lifts to $H^*(X^n)$ makes it clear that Theorem 1 still makes sense when X is a rational homology manifold. Indeed, the cohomology classes $\underline{z}$ and $\underline{a}$ are defined for any space satisfying Poincaré duality (i.e. with the global homological properties of an oriented differentiable manifold), while L(X) is defined for a rational homology manifold by the definition of Thom or of Milnor. In the spirit of Chapter I, we ask whether a result involving L-classes which is known to hold for differentiable manifolds and which involves no differentiable data (normal bundles, eigenvalues, etc.) also holds for rational homology manifolds. If we look at the proof of Theorem 1, we find that only at one place did we use the differentiability, namely an application of the G-signature theorem of Atiyah and Singer which was used to prove the following result:

Proposition 1: Let X be a compact, connected, closed, orientable differentiable manifold of dimension 2s. Let $\sigma: X^r \to X^r$ be the permutation map sending $(x_1, \ldots, x_r)$ to $(x_2, \ldots, x_r, x_1)$. Let $f: X^r \to S^{2k}$ be a smooth map with $f(\sigma \circ y) = f(y)$ for all $y \in X^r$, and choose $p \in S^{2k}$ as a regular value of $\underline{f}$ and of $f|_\Delta$ (where $\Delta \subset X^r$ is the diagonal); thus $A = f^{-1}(p)$ is a submanifold of X^r of dimension $2rs - 2k$ and $A \cap \Delta$ is a manifold of dimension $2r - 2k$. Then

$$\mathrm{Sign}(\sigma, A) = \begin{cases} e(X), & \text{if } \underline{r} \text{ is even, } k=0, \\ 0, & \text{if } \underline{r} \text{ is even, } k>0, \\ r^{-k}\,\mathrm{Sign}(A \cap \Delta), & \text{if } \underline{r} \text{ is odd.} \end{cases} \qquad (21)$$

Thus we have a statement, formulated purely in homological terms and with notions that make sense for X a rational homology manifold (we let $\underline{f}$ be simplicial and interpret transversality in the sense of Thom), but only known for X a smooth manifold. A proof of Proposition 1 for any rational homology manifold X would prove that Theorem 1 also holds for X, since the other steps in §§9-12 carry over without change. There is an analogous proposition if a finite group acts on X whose truth would imply the validity of the result of §14 for L(g,X(n)) for X a rational homology manifold.

We close this section with a discussion of equation (7) and its above-mentioned resemblance to a formula involving normal bundles. Recall (p. 1) that Thom showed how to define a bundle ν over a rational homology manifold embedded in a certain way (one says that N is a submanifold with orthogonal normal structure) in another rational homology manifold M. A rational homology manifold M always has enough such submanifolds to generate its rational homology. If M is a smooth manifold and N a smooth submanifold, then ν is the usual normal bundle, so we can consistently refer to ν in general as a normal bundle. It is a real vector bundle whose dimension equals the codimension of N in M. Finally, Thom showed that the L-classes of M and N (in his sense) are related to the L-class of ν just as in the differentiable case, i.e. by

$$j^*L(M) = L(\nu)L(N), \tag{22}$$

where $\underline{j}$ is the inclusion map from N to M.

Equations (7) and (22) are identical in form if we take for N and M the rational homology manifolds $X(n)$ and $X(n+1)$, and our first thought is that this formal identity cannot be accidental but rather arises because $X(n)$ really is a rational homology submanifold of $X(n+1)$ with orthogonal normal structure, its normal bundle ν_n in the sense of Thom then inevitably satisfying

$$L(\nu_n) = Q_s(\eta_n). \tag{23}$$

However, this is not the case if $\underline{s}$ is greater than one (if $s=1$, $X(n)$ is a differentiable and even a complex manifold, and ν_n is the normal bundle of $X(n)$ in $X(n+1)$ in the usual sense; the first Chern class of ν_n equals $\eta_n \in H^2(X(n))$ and $Q_s(\eta_n) = \eta_n/\tanh \eta_n$ is the usual L-class of a complex line bundle). To see this, we apply in reverse the multiplicative sequence used to define the L-class in terms of the Pontrjagin class (cf. Hirzebruch [12], §1, especially for the proof that the multiplicative sequence associated to $x/\tanh x$ is invertible). Since $Q_s(\eta)$ is a power series in η^2, it follows from the properties of multiplicative sequences that the sequence $R_s(\eta)$ obtained is also a power series in η^2. But η^2 has degree 4s, and the bundle ν_n (if it exists) has dimension 2s and therefore a Pontrjagin class cutting off at dimension 4s. Therefore if (23) holds, we have from $p(\nu_n) = R_s(\eta_n)$ that the series R_s must break off after the second term:

$$R_s(\eta) = 1 + \alpha\eta^2 . \tag{24}$$

But the multiplicative sequence $x/\tanh x$ applied to (24) gives as the corresponding L-class the series

$$1 + \beta_s \alpha\eta^2 + (\beta_s^2 - \beta_{2s})\alpha^2\eta^4/2 + \ldots , \tag{25}$$

where we have written β_k for the coefficient $2^{2k}(2^{2k-1}-1)B_k/(2k)!$ of x^{2k} in $x/\sin 2x$ (this coefficient is denoted by s_k in [12]). Since

$$Q_s(\eta) = 1 + (\frac{1}{3^s})\eta^2 + (\frac{1}{5^s} - \frac{2}{9^s})\eta^4 + \ldots, \tag{26}$$

it can only be of the form (25) if the relation holds which is obtained by equating coefficients of η^{2k} in (25) and (26) and eliminating α, i.e.

$$\frac{\beta_s^2 - \beta_{2s}}{2\beta_s^2} = \frac{9^s}{5^s} - 2. \tag{27}$$

It is easy to see that (27) is fulfilled only for $s=1$.

We have therefore proved that $X(n)$ is not in general a rational homology submanifold of $X(n+1)$ with orthogonal normal structure. However, the formal analogy between (7) and (22) still suggests that it has a particularly nice normal structure in some sense. We can hope to find some more general type of normal bundle than $O(n)$-bundles, for which (rational) Pontrjagin classes or L-classes are still defined and such that the inclusion of the nth symmetric product of a manifold in the $(n+1)^{st}$ always has a normal bundle in the generalised sense. In fact, a definition recently has been given for such a generalised bundle for homology manifolds (Martin and Maunder [26]). These objects, called "homology cobordism bundles," are (roughly) defined as projections $E \to B$ for which the inverse image of a cell in B is H-cobordant to the product of the cell with the fibre, the fibre being taken as D^n in an n-dimensional bundle. It was shown (in the paper referred to above and in Maunder [27]) that these objects form a reasonable category, allowing Whitney direct sums, that they possess an L-class, that the inclusion of N as a homology submanifold in a homology manifold M always has a normal homology cobordism bundle ν and (22) is satisfied, and finally that the set $K_H(X)$ of stable isomorphism classes of homology cobordism bundles over X is an abelian group and that $X \to K_H(X)$ is a representable functor. These results were proved in the category of homology manifolds but possibly still hold if one only requires that the spaces be rational homology manifolds, and replaces homology by rational homology in the definition of H-cobordism. Then equation (7) could be interpreted as saying that the L-class of the rational homology cobordism bundle ν_n of $X(n)$ in $X(n+1)$ is given by (23).

In [17], the Pontrjagin class of a homology cobordism bundle was defined by using Thom's definition of the rational Pontrjagin class of a homology manifold (as the class obtained by applying to the L-class defined in §2 the inverse of the Hirzebruch multiplicative sequence $x/\tanh x$) and then requiring (22) or its analogue for Pontrjagin classes to hold; this suffices to define $L(\nu)$ for any bundle since a bundle can be thought of as the normal bundle of its zero-section in its total space. However, this seems to be the wrong generalisation of the Pontrjagin class, since, as shown above, the Pontrjagin class defined in this way for ν_n does not break off at the right point. Indeed, it almost certainly does not break off at all, since the formal power series $Q_s(\eta)$ in a variable η of degree $2s$ does not seem to split as a finite product $\prod_i Q_1(x_i)$ with variables x_i of degree 2. It seems more appropriate to define the Pontrjagin class of our normal bundle ν_n by

$$p(\nu_n) = 1 + \eta_n^2. \tag{28}$$

If $s=1$, then ν_n is a complex line bundle over $X(n)$, and (28) is the usual Pontrjagin class. In general we call ν_n a "line bundle of type $\underline{s}$" (we omit in future the words "rational homology cobordism" before "bundle"). The standard model is obtained by taking for X a sphere of dimension $2s$. The fact that $Q_s(-)$ is independent of the number of factors $\underline{n}$ in the symmetric product suggests that the bundle ν_n over $S^{2s}(n)$ is itself independent of $\underline{n}$, i.e. that $j^*\nu_{n+1} = \nu_n$. In any case, this certainly holds for the L-class as defined in (23) or the Pontrjagin class as defined by (28). We then obtain as classifying space for line bundles of type $\underline{s}$ the limit $\varinjlim S^{2s}(n) = S^{2s}(\infty)$, over which there is a universal line bundle of type $\underline{s}$ defined as the limit of the bundles ν_n, and for this bundle ν we have

$$L(\nu) = Q_s(\eta), \qquad p(\nu) = 1 + \eta^2, \tag{29}$$

where η is the generator of $H^*(S^{2s}(\infty),\mathbb{Q}) = \mathbb{Q}[[\eta]]$. We then define in general a line bundle of type $\underline{s}$ to be the pull-back of ν to a space X by some map $f:X \to S^{2s}(\infty)$, its L-class and Pontrjagin class are then defined as the pull-backs of the elements (29). Therefore there is an isomorphism between the set of homotopy classes of maps $\underline{f}$ and the set of isomorphism classes of bundles of type $\underline{s}$ over X. But a well-known theorem of Dold and Thom [9] states that the homotopy groups of an infinite symmetric product $X(\infty)$ are isomorphic to the homology groups of X, and thus in particular that the space $S^{2s}(\infty)$ is a $K(\mathbb{Z},2s)$. The isomorphism classes of line bundles of type s are therefore in 1:1 correspondence with the elements of $[X,K(\mathbb{Z},2s)] \cong H^{2s}(X;\mathbb{Z})$. Thus a line bundle ξ over X is completely classified by a "first Chern class" $c_1(\xi) \in H^{2s}(X;\mathbb{Z})$, where $f:X \to S^{2s}(\infty)$ is the classifying map of ξ (i.e. $\xi \cong f^*\nu$).

All of these remarks are naturally conjectural only, depending on the existence of an appropriate category of rational homology bundles enjoying at least the properties of the Martin-Maunder bundles (namely that direct sums and induced bundles can be formed, that there is a classifying space, and that the bundles have L-classes). We can go further in the description of the form the theory might take if these bundles exist. For a direct sum $\xi = \oplus\xi_i$ of line bundles ξ_i of type s over a space X, we could define the total Chern, Pontrjagin and L-classes as $\Pi(1+x_i)$, $\Pi(1+x_i^2)$ and $\Pi Q_s(x_i)$, respectively, where $x_i = c_1(\xi_i)$. We might even hope that there is a splitting theorem analogous to the one for ordinary complex bundles, i.e. that there is a natural class of "bundles of type s" — bundles ξ such that $g^*\xi$ is a sum of line bundles, for some map $g:Y \to X$ for which $g^*:H^*(X) \to H^*(Y)$ is a monomorphism. Then if it is also true that the classes $c(g^*\xi)$, $p(g^*\xi)$ and $L(g^*\xi)$ defined above lie in the image of g^*, we have definitions for the corresponding characteristic classes for the bundle ξ. This would then suggest a whole series of further questions. First, we would want a geometric characterisation of those rational homology cobordism bundles which are of type s in this sense (a necessary condition, for example, would be that the L-class is zero in dimensions not divisible by 4s). This geometric description should be such as to allow the actual construction of the map g with $g^*\xi$ a sum of line bundles (analogous to the well-known construction for complex bundles). We could then ask if every bundle in our category is a direct sum of bundles of type s, and, if so, if the representation is unique (e.g. could it happen that a line bundle of type 6 is the direct sum of a line bundle of type 4 and one of type 2?). This question is clearly related to the independence of the various power series $Q_s(t)$. Knowing the answers to all of these questions would provide information about the cohomology of the classifying space for rational homology cobordism bundles. For instance, if the splitting into bundles of type $\underline{s}$ always exists and is unique, then the classifying space would contain elements p_i^s of degree $4is$ for every $i, s \geq 1$ (p_i^1 being the usual i^{th} Pontrjagin class). The classifying space for bundles of type $\underline{s}$ and dimension $\underline{n}$ (i.e. having a pull-back which is a sum of $\underline{n}$ line bundles of type $\underline{s}$) would be a space BU_n^s whose cohomology contains the classes p_i^s with $i \leq n$, and there would be a map $BU_n^s \leftarrow (K(\mathbb{Z}, 2s))^n$ (sending $\underline{n}$ line bundles over a space to their direct sum) which would map p_i^s to the ith elementary symmetric polynomial in $\eta_1^2, \ldots, \eta_n^2$ (where η_j is the generator of the cohomology of the jth factor $K(\mathbb{Z}, 2s)$). The L-class of a bundle with classifying map $f: X \to BU_n^s$ would have the component

$$L_i^s(f^*p_1^s, \ldots, f^*p_i^s) \in H^{4is}(X; \mathbb{Q}) \tag{30}$$

in degree $4is$ ($i \leq n$), where L_i^s is a generalized L-polynomial defined as the multiplicative sequence in the sense of Hirzebruch with characteristic power series $Q_s(\sqrt{t})$. The first few values of the polynomials L_i^s are given in the table on page 51, in the pious hope that some day there will be a theory to back up all these fancies.

TABLE: THE FUNCTION $Q_n(t)$ AND RELATED POWER SERIES

$$g_n(t) = t + \left(\frac{1}{3^n}\right)t^3 + \left(\frac{1}{5^n}\right)t^5 + \left(\frac{1}{7^n}\right)t^7 + \left(\frac{1}{9^n}\right)t^9 + \dots$$

$$f_n(t) = g_n^{-1}(t) = t + \left(-\frac{1}{3^n}\right)t^3 + \left(-\frac{1}{5^n} + \frac{3}{9^n}\right)t^5 +$$
$$+ \left(-\frac{1}{7^n} + \frac{8}{15^n} - \frac{12}{27^n}\right)t^7 + \left(-\frac{1}{9^n} + \frac{10}{21^n} + \frac{5}{25^n} - \frac{55}{45^n} + \frac{55}{81^n}\right)t^9 + \dots$$

$$Q_n(t) = \frac{t}{f_n(t)} = 1 + \left(\frac{1}{3^n}\right)t^2 + \left(\frac{1}{5^n} - \frac{2}{9^n}\right)t^4 + \left(\frac{1}{7^n} - \frac{6}{15^n} + \frac{7}{27^n}\right)t^6 +$$
$$+ \left(\frac{1}{9^n} - \frac{8}{21^n} - \frac{4}{25^n} + \frac{36}{45^n} - \frac{30}{81^n}\right)t^8 + \dots$$

$$\prod_{i=1}^{N} Q_n(x_i) = 1 + L_1^n(p_1) + L_2^n(p_1,p_2) + L_3^n(p_1,p_2,p_3) + \dots,$$

where $p_i = \sigma_i(x_1^2,\dots,x_N^2)$ (i^{th} elementary symmetric polynomial)

and:

$$L_1^n(p_1) = \left(\frac{1}{3^n}\right) p_1$$

$$L_2^n(p_1,p_2) = \left(\frac{1}{5^n} - \frac{2}{9^n}\right) p_1^2 + \left(-\frac{2}{5^n} + \frac{5}{9^n}\right) p_2$$

$$L_3^n(p_1,p_2,p_3) = \left(\frac{1}{7^n} - \frac{6}{15^n} + \frac{7}{27^n}\right) p_1^3 + \left(-\frac{3}{7^n} + \frac{19}{15^n} - \frac{23}{27^n}\right) p_1p_2 +$$
$$+ \left(\frac{3}{7^n} - \frac{21}{15^n} + \frac{28}{27^n}\right) p_3$$

$$L_4^n(p_1,p_2,p_3,p_4) = \left(\frac{1}{9^n} - \frac{8}{21^n} - \frac{4}{25^n} + \frac{36}{45^n} - \frac{30}{81^n}\right) p_1^4 +$$
$$+ \left(-\frac{4}{9^n} + \frac{33}{21^n} + \frac{16}{25^n} - \frac{150}{45^n} + \frac{127}{81^n}\right) p_1^2p_2$$
$$+ \left(\frac{4}{9^n} - \frac{33}{21^n} - \frac{18}{25^n} + \frac{159}{45^n} - \frac{137}{81^n}\right) p_1p_3$$
$$+ \left(\frac{2}{9^n} - \frac{18}{21^n} - \frac{7}{25^n} + \frac{80}{45^n} - \frac{70}{81^n}\right) p_2^2$$
$$+ \left(-\frac{4}{9^n} + \frac{36}{21^n} + \frac{18}{25^n} - \frac{180}{45^n} + \frac{165}{81^n}\right) p_4$$

§9. The action of S_n on X^n

We wish to calculate the L-class of $X(n) = X^n/S_n$, using the general formula for the L-class of a quotient space which was proved in §3. It is clear from that formula that the calculation consists of three steps: a calculation of the fixed-point sets $(X^n)^\sigma$ and their equivariant normal bundles (for all $\sigma \in S_n$) in order to compute $L'(\sigma, X^n) \in H^*((X^n)^\sigma)$; a description of the Gysin homomorphism of the inclusion $(X^n)^\sigma \subset X^n$ in order to evaluate $L(\sigma, X^n) \in H^*(X^n)$; and finally a summation over all $\sigma \in S_n$. These three steps will be carried out in this section and the two following ones, and will result in a formula which is already in a shape suitable for computations but in which the dependence of $L(X(n))$ on $\underline{n}$ has not yet been made explicit; a fourth section will then be devoted to the transformation of this expression into the one given in Section 8.

We begin, then, by examining the fixed-point set of the action of $\sigma \in S_n$ on X^n. The permutation $\underline{\sigma}$ can be written as a product of cycles, and it is clear from the definition of the action on X^n that this action has a corresponding decomposition as a product of actions of the form

$$\begin{array}{rccc} \sigma_r: & X^r & \rightarrow & X^r \\ & (x_1,\ldots,x_r) & \mapsto & (x_2,\ldots,x_r,x_1). \end{array} \tag{1}$$

Obviously for the standard action by cyclic permutation σ_r, the fixed-point set $(X^r)^{\sigma_r}$ consists of the points $(x,\ldots,x)$ for x X, i.e.

$$(X^r)^{\sigma_r} = \Delta_r \subset X^r \text{ (diagonal).} \tag{2}$$

The diagonal is naturally isomorphic to X under $x \leftrightarrow (x,\ldots,x)$; this isomorphism will be used tacitly in future, so that we shall consider the normal bundle of the fixed-point set of σ_r as a bundle over X and the inclusion of the fixed-point set as the diagonal map

$$\begin{array}{rccc} d = d_r: & X & \rightarrow & X^r, \\ & x & \mapsto & (x,\ldots,x). \end{array} \tag{3}$$

Returning to the permutation $\underline{\sigma}$ on X^n, we specify a little more precisely the decomposition into cyclic permutations. Let k_r be the number of cycles of length $\underline{r}$; thus k_1 is the number of integers $\underline{i}$ left fixed by $\underline{\sigma}$ and k_2 the number of pairs of integers $\underline{i}, \underline{j}$ interchanged by $\underline{\sigma}$.

Since $\underline{\sigma}$ acts on $\underline{n}$ integers altogether, we have

$$n = 1\cdot k_1 + 2\cdot k_2 + 3\cdot k_3 + \dots, \tag{4}$$

i.e. we have associated to the permutation $\underline{\sigma}$ of $\{1,\dots,n\}$ a partition $\pi = (k_1, k_2, \dots)$ of $\underline{n}$. Clearly the number of permutations with associated partition $\underline{\pi}$ is

$$N(\pi) = \frac{n!}{k_1!k_2!\dots 1^{k_1}2^{k_2}\dots}, \tag{5}$$

since of the n! ways of putting $\underline{n}$ objects into $k_1+k_2+\dots$ numbered boxes (there being k_r boxes having $\underline{r}$ slots, and the slots in each box also being numbered), two yield the same permutation if and only if there is a permutation of the k_r $\underline{r}$-boxes (giving the factor $k_r!$) or a cyclic permutation within an $\underline{r}$-box (giving a factor $\underline{r}$ for each of the k_r $\underline{r}$-boxes).

To illustrate the sort of calculation which must be done when working with these elements, we give another derivation of the formula of Macdonald for the Euler characteristic of X(n) which was proved earlier (Proposition 1 of §7) by a direct computation of the cohomology. We use the following formula for the Euler characteristic of the quotient of a space by a finite group action, which seems to be less well known than it should be:

$$e(X/G) = \frac{1}{|G|} \sum_{g\in G} e(X^g), \tag{6}$$

i.e. the desired Euler characteristic is just the average over G of the Euler characteristics of the fixed-point sets of the individual elements of G. [To see that (6) holds, we work (as usual) with rational coefficients, so that the cohomology of X/G is the G-invariant part of $H^*(X)$, and use the elementary result from linear algebra that the dimension of the G-invariant part of a G-vector space is the average over G of the traces of the individual elements of G. Then

$$e(X/G) = \sum_{i\geq 0} (-1)^i \dim H^i(X/G) = \sum_{i\geq 0} (-1)^i \dim H^i(X)^G$$

$$= \sum_{i\geq 0} (-1)^i \left(\frac{1}{|G|} \sum_{g\in G} \operatorname{tr}(g^*|H^i(X))\right) = \frac{1}{|G|} \sum_{g\in G} e(g,X),$$

where $e(g,X)$ is the equivariant Euler characteristic $\sum_i (-1)^i \operatorname{tr}(g^*|H^i(X))$. It only remains to show that

$$e(g,X) = e(X^g); \tag{7}$$

this is the Lefschetz index formula, and can also be immediately obtained from the equivariant Atiyah-Singer theorem on applying it to the de Rham complex of X (cf. [13], §9, eq. (11)).]

Applying (6) to our situation gives, since the fixed-point set of an element $\underline{\sigma}$ corresponding to the partition (4) has been found to be isomorphic to $X^{k_1+k_2+\cdots}$ and therefore to have Euler characteristic $e(X)^{k_1+k_2+\cdots}$,

$$n!\; e(X(n)) = \sum_{\pi \text{ a partition of } \underline{n}} N(\pi)\; e(X)^{k_1+k_2+\cdots}. \tag{8}$$

If we substitute expression (5) for $N(\pi)$ into this, we obtain

$$\sum_{n=0}^{\infty} t^n e(X(n)) = \sum_{k_1,k_2,\ldots\geq 0} \frac{t^{k_1+2k_2+3k_3+\cdots}\; e(X)^{k_1+k_2+\cdots}}{k_1!k_2!\cdots 1^{k_1}2^{k_2}\cdots}$$

$$= \prod_{r=1}^{\infty}\left(\sum_{k\geq 0} t^{rk} e(X)^k/r^k k!\right) = \prod_{r=1}^{\infty} e^{t^r e(X)/r} = e^{-e(X)\log(1-t)}$$

$$= (1-t)^{-e(X)}, \tag{9}$$

in agreement with equation (13) of §7. However, the method given here has the advantage, as well as that of illustrating the technique of manipulating averages over the symmetric group, that it can be used without change to compute the Euler characteristic of X^n/G for any subgroup G of S_n. For example, if we take $G = A_n$ to be the alternating group, the only change in (8) and (9) is that the sum is restricted to those k_r with $k_2+k_4+\ldots$ even (since we only have even permutations), and that the factor $|G|$ by which we have to divide is $n!/2$ rather than $n!$ (if $n\geq 2$). We deduce immediately

<u>Proposition 1</u>: For $n>1$, we have the equality

$$e(X^n/A_n) = \text{coefficient of } t^n \text{ in } \{(1-t)^{-e(X)} + (1+t)^{e(X)}\}. \tag{10}$$

We can apply exactly the same technique to the signature. Thus, if we set τ_r equal to the signature of the cyclic permutation (1),

$$\tau_r = \operatorname{Sign}(\sigma_r, X^r), \tag{11}$$

we see that the signature of a permutation σ with associated partition (4) is $\tau_1^{k_1}\tau_2^{k_2}\ldots$ (since signature behaves multiplicatively for a product action), and a computation exactly like (9) then yields

$$\sum_{n=0}^{\infty} t^n \operatorname{Sign}(X(n)) = \sum_{n=0}^{\infty} t^n \Big(\frac{1}{n!} \sum_{\sigma\in S_n} \operatorname{Sign}(\sigma, X^n) \Big)$$

$$= \exp\Big[\sum_{r=1}^{\infty} \tau_r t^r/r \Big] . \tag{12}$$

It will follow from the calculation of $L'(\sigma,X^n)$ later in this section that the integers τ_r are given by

$$\tau_r = \begin{cases} \tau(X), & \text{if } r \text{ is odd,} \\ e(X), & \text{if } r \text{ is even,} \end{cases} \tag{13}$$

where (X) is the signature of X, and therefore (12) becomes

<u>Theorem 1</u> (Hirzebruch [14]): The signature of the nth symmetric product $X(n)$ of an even-dimensional closed oriented manifold is given by

$$\sum_{n=0}^{\infty} t^n \operatorname{Sign}(X(n)) = \Big(\frac{1}{1-t^2} \Big)^{e(X)/2} \Big(\frac{1+t}{1-t} \Big)^{\operatorname{Sign}(X)/2} . \tag{14}$$

Note that the right-hand side of (14) is a rational function of t since $e(X)$ and $\operatorname{Sign}(X)$ are equal modulo 2.

We now turn to the main task of this section: the calculation of the action of σ on the normal bundle of $(X^n)^\sigma$ in X^n, and the resulting evaluation of $L'(\sigma,X^n)$. Since the equivariant L-classes behave multiplicatively for product actions, we can restrict our attention to the cyclic element σ_r of (1). As stated previously, we shall identify the fixed-point set Δ_r with X without explicit mention, so that when we speak of the normal bundle of the fixed-point set we mean its pull-back to X under this isomorphism; clearly this bundle is isomorphic to a sum of r-1 copies of the tangent bundle of X. At a point $(x,\ldots,x)$ of X^r, the tangent bundle of X^r consists of r-tuples $(v_1 \ldots, v_r)$ with $v_i \in T_xX$, and $T_x(\Delta_r)$ consists of r-tuples $(v,\ldots,v)$, so we can represent the fibre N_x of the normal bundle by these vectors with $v_1+\ldots+v_r=0$. The action of σ_r is given by cyclic permutation of the vectors v_i. Since σ_r has order r, its eigenvalues are $e^{2\pi ik/r}$, where k ranges from 1 to r-1. More precisely, for $k \neq r/2$, there is a subbundle $N(2\pi k/r)$ of N on which σ_r acts by the matrix

$$\begin{pmatrix} \cos 2\pi k/r & -\sin 2\pi k/r \\ \sin 2\pi k/r & \cos 2\pi k/r \end{pmatrix} \quad ; \tag{15}$$

the fibre $N_x(2\pi k/r)$ of this bundle at $(x,\ldots,x)$ consists of vectors $(v_1,\ldots,v_r)$ with $v_j = v' \cos 2\pi jk/r + v'' \sin 2\pi jk/r$ for some $v', v'' \in T_xX$, so that

$$N(2\pi k/r) \cong TX \oplus TX. \tag{16}$$

Since the matrix (15) is equivalent to the same matrix with k replaced by -k, we can assume $0 < k < r/2$. The remaining eigenvector -1 (in the case that r is even) has as eigenbundle the bundle $N(\pi)$ whose fibre at x consists of vectors $(v,-v,\ldots,-v)$, so

$$N(\pi) \cong TX. \tag{17}$$

The isomorphism (16) refers to $N(2\pi k/r)$ as a real bundle, but it also has the structure of a complex bundle if we require the action of σ_r to be given as multiplication by $e^{2\pi ik/r}$, and then

$$N(2\pi k/r) \cong TX \otimes \mathbb{C} . \tag{18}$$

We now have assembled sufficient information to apply the Atiyah-Singer formula, which, we recall, states

$$L'(g,Y) = \prod_{0<\theta<\pi} (i \tan \frac{\theta}{2})^{-\dim_{\mathbb{C}} N^g(\theta)} \cdot L(Y^g) \cdot L(N^g(\pi))^{-1} \cdot e(N^g(\pi)) \cdot \prod_{0<\theta<\pi} L_\theta(N^g(\theta)), \tag{19}$$

where $L_\theta(\xi)$ is a multiplicative sequence defined for complex bundles ξ by

$$L_\theta(\xi) = \prod_{x_j} \frac{\tanh i\theta/2}{\tanh (x_j + i\theta/2)} , \tag{20}$$

the x_j's being formal two-dimensional cohomology classes of the base space of ξ whose k^{th} elementary symmetric polynomial is $c_k(\xi)$.

We apply (19) with $Y = X^r$ and $g = \sigma_r$. First consider the case of even $\underline{r}$. Since $N^g(\pi) \cong TX$ is then a bundle of dimension equal to that of Y^g, the class (19) is only non-zero in the top dimension $2\underline{s}$. The various L- and L_θ-classes all have leading term 1, and therefore can be omitted. The Euler class of $N^g(\pi)$ is then $(-1)^{(r/2 - 1)s}\, e(X)\, z$, where $z \in H^{2s}(X)$ is the class of eq. (1) of § . The reason for the sign is that, in the Atiyah-Singer recipe, $N^g(\pi)$ is supposed to be oriented by the natural orientations on T(Y) and on the $N^g(\theta)$ for $\theta \neq \pi$ (assuming that Y^g also has a given orientation, which is the case here), the latter being given the orientation coming from their structure as complex bundles. Now it is known (see for instance [12], p. 66) that, if ξ is an oriented bundle of dimension $\underline{q}$, the orientation on $\xi \otimes \mathbb{C} \cong \xi \oplus \xi$ coming from the complex structure differs from the natural orientation on $\xi \oplus \xi$ by a factor $(-1)^{q(q-1)/2}$. Thus for $1 \leq k \leq (r-2)/2$, the orientation on $N^g(2\pi k/r)$ as a complex bundle differs from the orientation of $TX \oplus TX$ by a factor of $(-1)^s$, and therefore the orientations on $N^g(\pi)$ given by the Atiyah-Singer procedure and by its natural identification with TX differ by a factor $(-1)^{s(r-2)/2}$. If we put all this information into (19), and use

$$\dim_{\mathbb{C}} N^{\sigma_r}(2\pi k/r) = 2s \qquad (1 \leq k \leq \tfrac{r}{2} - 1), \tag{21}$$

we obtain

$$L'(\sigma_r, X^r) = \prod_{k=1}^{\frac{r}{2}-1} (i \tan \tfrac{\pi k}{r})^{-2s} \cdot (-1)^{s(r/2-1)} \cdot e(X) \cdot z, \tag{22}$$

or, since

$$(\tan \tfrac{\pi k}{r})\,(\tan \tfrac{\pi}{r}(\tfrac{r}{2} - k)) = 1, \tag{23}$$

$$L'(\sigma_r, X^r) = e(X)\, z \in H^{2s}(X) \qquad (r \text{ even}). \tag{24}$$

We now assume that $\underline{r}$ is odd. Again we can use (21) and (23) to see that the first product in (19) simply is the factor (this time $(-1)^{s(r-1)/2}$) giving the difference between the orientation of Y^g from the Atiyah-Singer recipe (i.e. induced from the orientations on Y and the complex bundles $N^g(\theta)$) and its orientation obtained by identifying it with X. Therefore (19) reduces to

$$L'(\sigma_r, X^r) = L(X) \prod_{k=1}^{\frac{r-1}{2}} L_{2\pi k/r}(N(2\pi k/r)). \tag{25}$$

Now, by definition, the Pontrjagin class of TX is $\Pi(1 + x_j^2)$ where

$c(TX\otimes\mathbb{C}) = \Pi(1 - x_j^2)$ (the Chern class of a bundle $\xi\otimes\mathbb{C}$ satisfies $2c_{2i+1} = 0$, and we are always working with rational or complex coefficients where torsion elements are killed). That is, by using the identification (18) we can take for the set of formal two-dimensional classes in (20) the set $\{x_j\} \cup \{-x_j\}$, where the x_j are the formal two-dimensional classes of X defined by

$$p(X) = \Pi(1 + x_j^2). \tag{26}$$

Therefore (20) and (25) combine to give

$$L'(\sigma_r, X^r) = \prod_{x_j} \left(\frac{x_j}{\tanh x_j} \prod_{k=1}^{(r-1)/2} \frac{\tanh i\pi k/r}{\tanh (x_j + i\pi k/r)} \frac{\tanh i\pi k/r}{\tanh(-x_j + i\pi k/r)} \right)$$

$$= \prod_{x_j} \frac{x_j}{\tanh rx_j} \qquad (r \text{ odd}), \tag{27}$$

where in the last line we have used the trigonometric identity

$$\prod_{k=-(r-1)/2}^{(r-1)/2} \coth (x + i\pi k/r) = \coth rx \quad (r \text{ odd}). \tag{28}$$

We state the results (24) and (27) together as a proposition:

<u>Proposition 2</u>: Let σ_r be the action (1) on X^r given by cyclic permutation of the coordinates. Then

$$L'(\sigma_r, X^r) = \begin{cases} e(X)\, z, & \text{if } r \text{ is even,} \\ \sum_{0 \leq c \leq s} r^{c-s} L_c(X), & \text{if } r \text{ is odd.} \end{cases} \tag{29}$$

Here $s = \frac{1}{2} \dim X$, $e(X)$ is the Euler characteristic, $\underline{z}$ is the element in $H^{2s}(X)$ given in (1) of §8, and $L_c(X)$ is the component in $H^{2c}(X)$ of the L-class of X.

An immediate corollary is that the number

$$\tau_r = \mathrm{Sign}(\sigma_r, X^r) = \langle L'(\sigma_r, X^r), [X^r] \rangle \tag{30}$$

has the value given in (13), completing the proof of Theorem 1.

§10. The Gysin homomorphism of the diagonal map

In this section we compute the Gysin homomorphism $d_!$ of the diagonal

$$\begin{aligned} d:\ X &\longrightarrow X^n, \\ x &\mapsto (x_1,\dots,x_n), \end{aligned} \tag{1}$$

which, we recall, is defined by

$$d_! = D_{X^n}^{-1}\, d_*\, D_X :\ H^*(X) \longrightarrow H^*(X^n), \tag{2}$$

where D_X is the Poincaré duality isomorphism from cohomology to homology defined by capping with the fundamental class, and d_* is the map induced by $\underline{d}$ in homology.

All notation will be as in previous sections; thus X is a closed oriented manifold of even dimension and two bases $\{e_i\}_{i\in I}$ and $\{f_i\}_{i\in I}$ for $H^*(X)$ are given which are dual to each other under the intersection pairing. All that will be used are elementary properties of the various products (cup, cap, slant) in homology and cohomology, as given, for instance, in 5.6, 6.1, and 6.10 of Spanier.

We first give a formulation of the result in terms of the given bases for $H^*(X)$; afterwards we will restate it in a basis-free manner.

Proposition 1: For $x \in H^*(X)$, the following formula holds:

$$d_! x = \sum_{i_1,\dots,i_n\in I} \varepsilon_{i_1\dots i_n} \langle e_{i_1}\cdots e_{i_n} x, [X]\rangle\, f_{i_1}\times\dots\times f_{i_n}, \tag{3}$$

where $\varepsilon_{i_1\dots i_n} = \pm 1$ is the sign obtained on rearranging $e_{i_1}\cdots e_{i_n} f_{i_1}\cdots f_{i_n}$ as $\pm e_{i_1} f_{i_1}\cdots e_{i_n} f_{i_n}$ and taking into account graded commutativity (this sign is equal to $(-1)^{r(r-1)/2}$, where $\underline{r}$ is the number of e_{i_j} of odd degree.)

Proof: Since $H^*(X^n)$ is spanned by the elements $e_{j_1}\times\dots\times e_{j_n}$ $(j_1,\dots,j_n\in I)$, it is sufficient to show that the two sides of (3) agree after we cup with this element on the left and evaluate on $[X^n]$. Since $\langle e_i f_j, [X]\rangle = \delta_{ij}$,

$$\langle (e_{j_1}\times\dots\times e_{j_n})(f_{i_1}\times\dots\times f_{i_n}), [X^n]\rangle$$

$$= \begin{cases} 0, & \text{if } (i_1,\dots,i_n) \neq (j_1,\dots,j_n), \\ \varepsilon_{j_1\dots j_n}, & \text{if } (i_1,\dots,i_n) = (j_1,\dots,j_n). \end{cases} \tag{4}$$

Therefore it is only necessary to prove

$$<(e_{j_1}\times\ldots\times e_{j_n})(d_!x),[X^n]> \;=\; <e_{j_1}\ldots e_{j_n}x,[X]> . \tag{5}$$

But from the definition of $d_!$ and elementary manipulations of the cup and cap products, we obtain

$$\begin{aligned} <(e_{j_1}\times\ldots\times e_{j_n})(d_!x),[X^n]> &= <e_{j_1}\times\ldots\times e_{j_n}, (d_!x)\cap[X^n]> \\ &= <e_{j_1}\times\ldots\times e_{j_n}, D_{X^n}(d_!x)> \\ &= <e_{j_1}\times\ldots\times e_{j_n}, d_*(D_X x)> \\ &= <d^*(e_{j_1}\times\ldots\times e_{j_n}), x\cap[X]> \\ &= <d^*(e_{j_1}\times\ldots\times e_{j_n})x,[X]> \end{aligned} \tag{6}$$

But $d^*(e_{j_1}\times\ldots\times e_{j_n}) = e_{j_1}\cup\ldots\cup e_{j_n}$ by definition of the cup product.

We now wish to reformulate the content of the proposition in a basis-free way. To do this, we use the slant product. Recall that, for two topological spaces A and B, the slant product sends an element $u\in H^r(A\times B)$ and an element $z\in H_q(B)$ to $u/z\in H^{r-q}(A)$. If $\underline{u}$ is a product $a\times b$ (with $a\in H^*(A)$, $b\in H^*(B)$), we have the formula

$$(a\times b)/z \;=\; <b,z>\,a \;\in\; H^*(A). \tag{7}$$

Now take $A = X^n$, $B = X$, and $z = [X]\in H_{2s}(X)$. Then

$$(f_{i_1}\times\ldots\times f_{i_n}\times y)/[X] \;=\; <y,[X]>\,f_{i_1}\times\ldots\times f_{i_n} \;\in\; H^*(X^n) \tag{8}$$

for all $y\in H^*(X)$. Substituting this into (3) and using graded commutativity, we obtain:

$$\begin{aligned} d_!x &= \sum_{i_1,\ldots,i_n\in I} \varepsilon_{i_1\ldots i_n}\{(f_{i_1}\times\ldots\times f_{i_n})\times(e_{i_1}\ldots e_{i_n}x)\}/[X] \\ &= \Sigma\ \varepsilon_{i_1\ldots i_n}\{(\pi_1^*f_{i_1})\ldots(\pi_n^*f_{i_n})(e_{i_1}\ldots e_{i_n}x)\}/[X] \\ &= \Sigma\ \{(\pi_1^*f_{i_1}\times e_{i_1})(\pi_2^*f_{i_2}\times e_{i_2})\ldots(\pi_n^*f_{i_n}\times e_{i_n})(1\times x)\}/[X], \end{aligned}$$

where the summation is always over the same indices. Here π_j is the projection $X^n\to X$ onto the jth factor, and the expression in curly brackets is an element of $H^*(X^n\times X)$. In the last line we can bring the summation into the product, obtaining

$$d_! x = \{(\sum_{i\in I} \pi_1^* f_i \times e_i)\ldots(\sum_{i\in I} \pi_n^* f_i \times e_i)(1\times x)\}/[X]. \tag{9}$$

Clearly

$$\sum_{i\in I} \pi_j^* f_i \times e_i = (\pi_j \times 1)^* a \in H^*(X^n \times X), \tag{10}$$

where

$$a = \sum_{i\in I} f_i \times e_i \in H^{2s}(X\times X) \tag{11}$$

and where

$$\pi_j \times 1 : X^n \times X \to X \times X \tag{12}$$

is the product of the jth projection and the identity map. The element $\underline{a}$ defined by (11) is independent of the choice of bases $\{e_i\}$ and $\{f_i\}$. One way to see this is to notice that, if $e_i' = \Sigma\, c_{ij} e_j$ and $f_i' = \Sigma\, d_{ij} f_j$ are another pair of dual bases, then the matrix $\underline{c}$ is the transpose of the inverse of $\underline{d}$, from which it follows that $\Sigma\, f_i' \times e_i' = \Sigma\, f_i \times e_i$. Another proof is to notice that, by the case n=2 of Proposition 1, we have

$$a = d_! 1 \in H^{2s}(X\times X), \tag{13}$$

where $d: X \to X\times X$ is the diagonal. Yet a third way is to deduce from equation (11) the identity (for all $x,y \in H^*(X)$)

$$\langle (x\times y)a, [X\times X] \rangle = \langle xy, [X] \rangle \tag{14}$$

and to notice that this characterizes $\underline{a}$ completely. Finally, from Lemma 6.10.1 of Spanier we see that $\underline{a}$ is the restriction to $X\times X$ of the Thom class

$$U \in H^{2s}(X\times X, X\times X - d(X)) \tag{15}$$

whose existence is equivalent to the orientability of X.

We can now restate Proposition 1 as:

osition 2: Let $d: X \to X^n$ be the diagonal map, $\pi_j : X^n \to X$ the projection the jth factor, $\pi_j \times 1 : X^n \times X \to X\times X$ the product of π_j with id_X, and $a \in H^{2s}(X\times X)$ the element defined by (13) or (14). Then for all $x \in H^*(X)$, the image of $\underline{x}$ under the Gysin homomorphism $d_!$ of $\underline{d}$ is given by the formula

$$d_! x = \{((\pi_1 \times 1)^* a) \ldots ((\pi_n \times 1)^* a)(1 \times x)\}/[X] \in H^*(X^n). \qquad (16)$$

§11. Preliminary formula for $L(X(n))$

We can now apply the result of Chapter I, which here states

$$L(X(n)) = \sum_{\sigma \in S_n} L(\sigma, X^n) \; ; \tag{1}$$

we omit the π^* since we are already using it to identify $H^*(X(n))$ with $H^*(X^n)^{S_n}$. If $\underline{\sigma}$ is broken up into a product of cyclic permutations σ_r then $L(\sigma, X^n)$ is the $\times$-product of the corresponding $L(\sigma_r, X^r)$. The latter are obtained by combining the results of the preceding two sections, and are given by

$$L(\sigma_r, X^r) = (d_r)_! L'(\sigma_r, X^r)$$

$$= \begin{cases} e(X)\; z \times \dots \times z, & \text{if } r \text{ is even,} \\ \sum\limits_{0 \leq c \leq s} r^{c-s} \{((\pi_1 \times 1)^* a) \dots ((\pi_r \times 1)^* a)(1 \times L_c(X))\}/[X], & \text{if } r \text{ is odd.} \end{cases} \tag{2}$$

For the first line we do not need the results of the last section, since by definition of the Gysin homomorphism and of the element z_X of the top-dimensional cohomology group of a manifold X we have $z_Y = f_* z_X$ for any map $f: X \to Y$ at all.

Now if $(j_1, \dots, j_r)$ is the set of integers cyclically permuted by $\underline{\sigma}$, then the ith projection map $\pi_i : X^r \to X$ is replaced by $\pi_{j_i} : X^n \to X$ when we identify the X^r on which σ_r acts with the $X_{j_1} \times \dots \times X_{j_r}$ on which $\underline{\sigma}$ acts (here X_j denotes the jth factor of X^n). This substitution must thus be made in (2) to obtain the factor of $L(\sigma, X^n)$ corresponding to the factor σ_r of $\underline{\sigma}$; for example, for even $\underline{r}$ we obtain $e(X)\,(\pi_{j_1}^* z) \dots (\pi_{j_r}^* z)$. In principle we have to keep track of the order of the j_i's (determined up to cyclic permutation by $\underline{\sigma}$), but since the cohomology classes $\underline{a}$, $\underline{z}$, and $L_c(X)$ are all of even degree and thus commute with everything, we will not in fact have to do so. Thus, for $A = \{j_1, \dots, j_r\}$ any subset of $N = \{1, \dots, n\}$ we have defined a class in $H^*(X^n)$

$$L(A) = \begin{cases} e(X) \prod_{i \in A} \pi_i^* z, & \text{if } r = |A| \text{ is even,} \\ \sum\limits_{0 \leq c \leq s} r^{c-s} \{\prod\limits_{i \in A} ((\pi_i \times 1)^* a) \cdot (1 \times L_c(X))\}/[X], & \text{if } r \text{ is odd,} \end{cases} \tag{3}$$

and for any $\sigma \in S_n$ we have

$$L(\sigma,X^n) = \prod_{A \text{ a cycle of } \sigma} L(A). \tag{4}$$

We repeat that the class $L(A)$ is even-dimensional and independent of the ordering of the elements of A.

When we calculated the Euler characteristic and signature of $X(n)$ in the last section, we wrote the summation over the symmetric group S_n as a summation over all partitions $\underline{\pi}$ of $\underline{n}$, followed by a summation over the $N(\pi)$ permutations in S_n with associated partition $\underline{\pi}$. In a similar way, we now write the summation occurring in (1) as a sum over all partitions of N into subsets A_i followed by a sum over all permutations of N whose cycles are precisely the A_i. Since (4) tells us that $L(\sigma,X^n)$ then only depends on the A_i's, the latter sum will simply be $\Pi_i L(A_i)$ times the number $N(A_1,\ldots,A_k)$ of permutations of N consisting precisely of cyclic permutations of each A_i. A set of $\underline{r}$ elements clearly has $(r-1)!$ cyclic permutations, so

$$N(A_1,\ldots,A_k) = (|A_1| - 1)! \ldots (|A_k| - 1)! . \tag{5}$$

We also have to divide the sum over all subsets $A_1,\ldots,A_k$ by $k!$, since the same partition of N into disjoint subsets is counted $k!$ times with different numberings of the A_i's. Therefore we can write equation (1) as

$$L(X(n)) = \sum_{k=1}^{n} \sum_{\substack{A_1,\ldots,A_k \subset N \\ A_i\text{'s disjoint} \\ A_1\cup\ldots\cup A_k = N}} \frac{1}{k!} \prod_{j=1}^{k} \Big((|A_j| - 1)!\; L(A_j) \Big) , \tag{6}$$

which, combined with equation (3) for $L(A)$, completes the determination of the L-class of $X(n)$. Nevertheless, the expression obtained is extremely unwieldy, and we will devote the remainder of this section and the whole of the next one to a reformulation of it into a better form.

We introduce dummy variables $x_1,\ldots,x_n$; the function of x_j will be as a marker, indicating the jth factor in the product X^n or its cohomology $H(X)\otimes\ldots\otimes H(X_n)$. The x_j's are supposed to commute with one another and with the elements of the cohomology of X^n but to satisfy no other relations; i.e. we will be working in the ring $H^*(X^n)[x_1,\ldots,x_n]$ of polynomials in the x_j's with coefficients in $H^*(X^n)$. We fix the following notations, which will be used throughout this and the next sections: N will always refer to the set $\{1,\dots,n\}$; for $A \subset N$, we write x_A for $\prod_{i\in A} x_i$; for $\underline{f}$ a power series or polynomial in the x_i's and A a subset of N, $\partial_A f$ is the coefficient of x_A in $\underline{f}$. Then for $A_1,\dots,A_k$ arbitrary subsets of N, the conditions that $1 \leq k \leq n$, that the sets A_i are disjoint, and that the union of the A_i's is N, are together equivalent to the single condition $x_{A_1}\dots x_{A_k} = x_N$. Therefore (6) becomes

$$L(X(n)) = \sum_{k=0}^{\infty} \frac{1}{k!} \sum_{\substack{A_1,\dots,A_k \subset N \\ x_{A_1}\dots x_{A_k} = x_N}} \prod_{j=1}^{k} \Big((|A_j| - 1)!\, L(A_j) \Big)$$

$$= \partial_N \left(\sum_{k=0}^{\infty} \frac{1}{k!} \sum_{A_1,\dots,A_k \subset N} \prod_{j=1}^{k} \Big((|A_j|-1)!\, x_{A_j}\, L(A_j) \Big) \right)$$

$$= \partial_N \left(\sum_{k=0}^{\infty} \frac{1}{k!} \prod_{j=1}^{k} \Big(\sum_{A\subset N} (|A|-1)!\, x_A\, L(A) \Big) \right)$$

$$= \partial_N \Big[\exp\Big(\sum_{A\subset N} (|A|-1)!\, x_A\, L(A) \Big)\Big] . \tag{7}$$

This expression, equivalent to (6), is considerably more pleasing. To make further progress, we must substitute the value of L(A) from (3). We write the exponent in (7) as $K(x_1,\dots,x_n)$ or simply K, and the corresponding sum restricted to subsets A with exactly $\underline{r}$ elements as K_r. Thus for $\underline{r}$ even, we obtain

$$K_r = \sum_{\substack{A\subset N \\ |A|=r}} (r-1)!\, x_A\, e(X)\, z_A \tag{8}$$

(we use a similar notation to that for x_A, i.e. $z_A = \prod_{i\in A} z_i$, where z_i denotes $\pi_i^* z = 1\times\dots\times z\times\dots\times 1 \in H^*(X^n)$, the product having a $\underline{z}$ in the ith place). Since the summation over A is a sum over unordered subsets of N, we have to divide by r! if we sum instead over all $i_1,\dots,i_r \in N$, so (8) becomes

$$K_r = \sum_{i_1,\dots,i_r \in N} \frac{1}{r}\, e(X)\, x_{i_1}\dots x_{i_r}\, z_{i_1}\dots z_{i_r}$$

$$= \sum_{i_1=1}^{n} \cdots \sum_{i_r=1}^{n} \frac{e(X)}{r} x_{i_1} z_{i_1} \cdots x_{i_r} z_{i_r}$$

$$= \frac{e(X)}{r} \prod_{j=1}^{r} \left(\sum_{i=1}^{n} x_i z_i \right)$$

$$= e(X) \ (x_1 z_1 + \dots + x_n z_n)^r / r \ , \qquad (r \text{ even}). \tag{9}$$

Exactly similarly, if $\underline{r}$ is odd we obtain from the second line of (3):

$$K_r = \frac{1}{r} \sum_{c=0}^{s} r^{c-s} \{(x_1(\pi_1 \times 1)^* a + \dots + x_n(\pi_n \times 1)^* a)^r (1 \times L_c(X))\}/[X], \tag{10}$$

where the expression in curly brackets is in the ring

$$R' = H^*(X^n \times X)[x_1, \dots, x_n] \tag{11}$$

which is mapped by $/[X]$ to the ring

$$R = H^*(X^n)[x_1, \dots, x_n]. \tag{12}$$

More precisely, the expressions considered are in the S_n-invariant subrings R'^{S_n} and R^{S_n}, where S_n acts by simultaneously permuting the factors of the tensor product $H^*(X^n) = \otimes_{i=1}^{n} H^*(X)$ and the x_i's.

We write

$$\bar{z} = x_1 z_1 + \dots + x_n z_n \in R, \tag{13}$$

$$\bar{a} = x_1(\pi_1 \times 1)^* a + \dots + x_n(\pi_n \times 1)^* a \in R' \tag{14}$$

for the elements appearing in equations (9) and (10). Then the exponent K of (8) is given by

$$K = \sum_{r=1}^{\infty} K_r = \sum_{\substack{r=1 \\ r \text{ even}}}^{\infty} e(X) \ \bar{z}^r / r + \sum_{\substack{r=1 \\ r \text{ odd}}}^{\infty} \sum_{c=0}^{s} r^{c-s-1} \{ \bar{a}^r (1 \times L_c(X))\}/[X]$$

$$= \frac{-e(X)}{2} \log(1 - \bar{z}^2) + \sum_{c=0}^{s} \{g_{s+1-c}(\bar{a}) \ (1 \times L_c(X))\}/[X], \tag{15}$$

where the function g_{s+1-c} is the power series defined in §8. This K is an element of the ring R defined in (12), e^K therefore also is, the symbol ∂_N defined above maps R to $H^*(X^n)$, and our final result is

$$L(X(n)) = \partial_N(e^K) \in H^*(X^n)^{S_n}. \tag{16}$$

§12. The dependence of L(X(n)) on n

In this section we investigate the dependence on n of the expression for L(X(n)) found in the last section, and complete the proof of Theorem 1 of §8.

We continue to use the notations of the last section involving the dummy variables x_i. The elements of $R = H^*(X^n)[x_1,\dots,x_n]$ which occurred were all polynomials in elements of the form

$$x_1\ \pi_1^*f + \dots + x_n\ \pi_n^*f \quad \in \quad R^{S_n} \subset R, \quad (f \in H^*(X)). \tag{1}$$

With the notation $f_0=z,\dots,f_b=1$ for a basis of $H^*(X)$ used in the preceding questions, the element (1) can be written as a linear combination of the elements

$$t_i \;=\; x_1\ \pi_1^*f_i + \dots + x_n\ \pi_n^*f_i \quad \in \quad R^{S_n} \quad (0 \leq i \leq b), \tag{2}$$

so that we are really only interested in the subring $\mathbb{Q}[t_0,\dots,t_b]$ of R^{S_n}. Note that $t_0 = \bar{z}$ and $t_b = x_1+\dots+x_n$. These t_i are the t_i of the theorem of §8, which were there introduced as rather mysterious dummy variables with the same commutation properties as the f_i, but which now are defined more naturally in terms of ordinary scalar or commuting dummy variables $x_1,\dots,x_n$. To express $\bar{a} \in R'$ in terms of the t_i, we use formula (11) of §10 to write

$$\bar{a} \;=\; \sum_{i=0}^{b} t_i \times e_i \quad \in \quad \mathbb{Q}[t_0,\dots,t_b] \otimes H^*(X) \subset R \otimes H^*(X) \cong R'. \tag{3}$$

Then the main result (15), (16) of §11 is

$$L(X(n)) = \partial_N\ G(t_0,\dots,t_b), \tag{4}$$

where

$$G(t_0,\dots,t_b) = \left[\frac{1}{(1-t_0^2)^{e(X)/2}}\right] e^{\sum_{c=0}^{s} g_{s-c+1}(\Sigma\, t_i \times e_i)(1\times L_c(X))/[X]} . \tag{5}$$

We now look at the effect of the restriction $j^*: H^*(X^{n+1}) \to H^*(X^n)$. We label all objects in the cohomology of X^n with a subscript $\underline{n}$. Then it follows immediately from the description of j^* (eq. (15) of §7) that

$$j^*(\ (t_i)_{n+1}\ ) \;=\; \begin{cases} (t_i)_n\,, & \text{if } 0 \leq i < b, \\ x_1 + \dots + x_{n+1}, & \text{if } i = b. \end{cases} \tag{6}$$

As a consequence, we find that, for $B \subset N$, and any polynomial F,

$$j^*\{\partial_B[F((t_0)_{n+1},\ldots,(t_b)_{n+1})]\} = \partial_B[F((t_0)_n,\ldots,(t_b)_n)] \in H^*(X^n). \quad (7)$$

The reason that we cannot use (7) to assert the stability under j^* of the expression (4) is that the set N itself changes as we pass from $n+1$ to $\underline{n}$.

It is clear from (6) that to study the effect of j^* on a polynomial F in $t_0,\ldots,t_b$ it is necessary to look separately at the dependence of F on the first $\underline{b}$ variables $t_0,\ldots,t_{b-1}$ and on the last variable $t_b = x_1+\ldots+x_n$. The former relates to the stable elements $[n_0(f_0)\ldots n_{b-1}(f_{b-1})] \in H^*(X(n))$ defined in eq. (17) of §7, namely

$$\sum_{B\subset N} \partial_B(t_0^{n_0}\ldots t_{b-1}^{n_{b-1}}) = [n_0(f_0)\ldots n_{b-1}(f_{b-1})] \in H^*(X^n). \quad (8)$$

To see that (8) holds, set $r = n_0+\ldots+n_{b-1}$ and note that the summands on the left-hand side of (8) with $|B| \neq r$ are certainly zero. Thus (replacing $t_0^{n_0}\ldots t_{b-1}^{n_{b-1}}$ by $t_{j_1}\ldots t_{j_r}$, where $j_1,\ldots,j_r \in \{0,\ldots,b-1\}$ are indices, not necessarily distinct) the left-hand side of (8) is

$$\sum_{1\leq i_1<\ldots<i_r\leq n} \partial_{\{i_1,\ldots,i_r\}}(t_{j_1}\ldots t_{j_r}),$$

i.e. we expand $t_{j_1}\ldots t_{j_r}$ as a polynomial in the x_i's, omit any term in which some x_i appears to a higher power than the first, and then set all the x_i's equal to one. Clearly this yields $\frac{1}{(n-r)!}\langle t_{j_1},\ldots, t_{j_r},1,\ldots,1\rangle$ in the notation of (8) of §7, where the number of 1's is $n-r$. Re-expressing this in terms of the notation of (9) and (17) of §7 then gives equation (8).

It remains to study the effect on (4) of the dependence of G on the variable t_b. First we note that since the slant product with $[X]$ sends $u\times v$ to $u\langle v,[X]\rangle$ and since $e_b=z$ has the same dimension as $[X]$, the only non-zero monomials of the form

$$\{(t_0\times e_0)^{k_0}\ldots(t_b\times e_b)^{k_b}(1\times L_c(X))\}/[X] \qquad (k_b > 0) \quad (9)$$

are those with $k_1=\ldots=k_{b-1}=0$, $c=0$, and $k_b=1$ (for which values (9) equals $t_b t_0^{k_0}$). Therefore an expansion of the exponent in (5) as a Taylor series in powers of $(t_b\times e_b)$ consists of the constant term obtained by setting t_b equal to zero, plus a linear term $t_b g'_{s+1}(t_0)$:

$$G(t_0,\dots,t_b) \quad = \quad e^{t_b g'_{s+1}(t_0)/t_0}\, G(t_0,\dots,t_{b-1},0). \tag{10}$$

But $g'_{s+1}(t) = g_s(t)/t$ (this follows immediately from the definition of g_s), and the value of $G(t_0,\dots,t_{b-1},0)$ obtained from (5) is exactly the power series of eq. (17) of Theorem 1 in §8, so we find that the only step still necessary to deduce that theorem from (4) is the following fact about power series.

<u>Proposition 1</u>: Let $G = G(t_0,\dots,t_{b-1})$ be a power series, g_s the series defined in (11) of §8, and f_s and Q_s the power series defined by

$$f_s(t) = g_s^{-1}(t) = t - \dots, \qquad Q_s(t) = t/f_s(t) = 1 + \dots . \tag{11}$$

Then

$$\partial_N\left(e^{(x_1+\dots+x_n)\, g_s(t_0)/t_0}\, G \right) \quad = \quad f'_s(\eta)Q_s(\eta)^{n+1} \sum_{A\subset N} Q_s(\eta)^{-|A|}\partial_A G. \tag{12}$$

Here f'_s denotes the first derivative, and $\eta = z_1+\dots+z_n$.

<u>Proof</u>: We write ξ for $x_1+\dots+x_n$ and $\varphi = \varphi(t_0)$ for $g_s(t_0)/t_0$. The definition of ∂_A as the operator sending a power series <u>f</u> to the coefficient of x_A in <u>f</u> is clearly equivalent to the formula

$$\partial_A f \quad = \quad \left(\frac{\partial^r}{\partial x_{i_1}\dots\partial x_{i_r}} f \right)_{x_1=\dots=x_n=0} \tag{13}$$

(this is the reason for the notation ∂_A), and therefore we can make use of a generalized Leibniz's rule for ∂_A of a product of two functions:

$$\partial_A(F_1F_2) \quad = \sum_{\substack{B,C\subset A\\ B\cup C=A\\ B\cap C=0}} \partial_B(F_1)\,\partial_C(F_2) \quad = \quad \sum_{B\subset A} \partial_B(F_1)\partial_{A-B}(F_2). \tag{14}$$

We apply this rule to the product on the left-hand side of (12) to get

$$\partial_N(\, e^{\xi\varphi}\, G\,) = \sum_{A\subset N} \partial_A G\cdot\partial_{N-A}(e^{\xi\varphi}) . \tag{15}$$

Now we expand the exponential as a power series and apply (14) again:

$$\partial_N(\, e^{\xi\varphi}\, G\,) = \sum_{A\subset N} \partial_A G \sum_{r=0}^{\infty} \frac{1}{r!} \sum_{B\subset N-A} \partial_B(\varphi^r)\partial_{N-A-B}(\xi^r). \tag{16}$$

But $\xi = x_1 + \ldots + x_r$, so $\xi^r = \sum_{i_1,\ldots,i_r \in N} x_{i_1} \ldots x_{i_r}$, and therefore

$$\partial_{N-A-B}(\xi^r) = \begin{cases} r!, & \text{if } |N-A-B| = r, \\ 0, & \text{if } |N-A-B| \neq r, \end{cases} \tag{17}$$

and (16) simplifies to

$$\sum_{\substack{A,B\subset N \\ A\cap B=\emptyset}} \partial_A G \; \partial_B(\varphi^{|N-A-B|}). \tag{18}$$

Now for any function $\underline{h}$ of one variable, and $C \subset N$, we have

$$\begin{aligned} \partial_C(h(t_0)) &= \text{coefficient of } x_C \text{ in } h(x_1 z_1 + \ldots + x_n z_n) \\ &= (|C|)!\, z_C \cdot \text{coefficient of } t^{|C|} \text{ in } h(t) \\ &= z_C \cdot (d^{|C|}/dt^{|C|}) h(t)|_{t=0}, \end{aligned} \tag{19}$$

where z_C denotes $\Pi_{i\in C}\, z_i$. Moreover, since G is a function only of $t_0,\ldots,t_{b-1}$, and since $f_i z = 0$ for $i=0,\ldots,b-1$ (here we need that f_i is of positive degree, i.e. that G is independent of t_b), we conclude that $z_B\, \partial_A G = 0$ if $A\cap B \neq \emptyset$ (for if $\underline{j}$ is in both A and B, then in the jth place in a typical monomial we have both a factor $\underline{z}$ and a factor f_i for some $i=0,\ldots,b-1$, and their product is zero). Therefore we can substitute (19) in (18) and omit the condition $A\cap B=\emptyset$:

$$\partial_N(e^{\xi\varphi} G) = \sum_{A\subset N} \partial_A G \sum_{B\subset N} z_B \cdot \frac{d^{|B|}}{dt^{|B|}}\left[\left(\frac{g_s(t)}{t}\right)^{n-|A|-|B|}\right]_{t=0}. \tag{20}$$

Comparing this with the equation (12) which we want to prove, we see that it only remains to prove, for $0\leq j\leq n$, the formula

$$\sum_{k=0}^{n}\left[\sum_{\substack{B\subset N \\ |B|=k}} z_B\right] \frac{d^k}{dt^k}\left[\left(\frac{g_s(t)}{t}\right)^{n-j-k}\right]_{t=0} = Q_s(\eta)^{n+1-j}\, f_s'(\eta). \tag{21}$$

But it follows easily from $z^2 = 0$ that the first expression in square brackets in (21) is just $\eta^k/k!$, while the second factor equals

$$k!\ \mathrm{res}_{t=0}[(g_s(t)/t)^{n-j-k}\, t^{-k-1}\, dt] = k!\ \mathrm{res}_{y=0}[Q_s(y)^{n-j+1} y^{-k-1} f_s'(y)dy],$$

as we see by substituting $y = g_s(t)$, $t = f_s(y)$. Equation (21) follows.

§13. Symmetric products of spheres and of Riemann surfaces

This section contains three applications of the theorem of §8. The first, included solely for the purpose of illustrating the use of that theorem, is a second derivation of Hirzebruch's formula for the signature of $X(n)$ (§9 eq. (14)). The second is an evaluation of the formula for X an even-dimensional sphere; the calculation here is very easy since the cohomology and L-class of X are trivial. The third application, rather harder, is to the case where X is two-dimensional, i.e. a Riemann surface (the case where $Q_s(t)$ is the usual series $t/\tanh t$). In this case, $X(n)$ is a complex manifold, a projective fibre bundle over the Jacobian of X if $n+e(X)>0$, and its Chern class has been calculated (Macdonald [15]). This of course also gives the L-class of X, and therefore we are able to check the correctness of the main theorem of this chapter.

We start, then, by evaluating $\mathrm{Sign}(X(n))$. Since $z^2=0$, we have

$$\eta^n/n! \;=\; z_1\ldots z_n \tag{1}$$

(a similar formula for all $\eta^k/k!$ was already used at the end of the last section). Therefore

$$\begin{aligned}
\mathrm{Sign}(X(n)) &= \langle L(X(n)),[X(n)]\rangle \\
&= (\deg \pi)^{-1}\langle L(X(n)),\pi_*[X^n]\rangle \\
&= \frac{1}{n!}\langle \pi^*L(X(n)),[X^n]\rangle \\
&= \frac{1}{n!}\ \text{coefficient of } z_1\ldots z_n \text{ in } \pi^*L(X(n)) \\
&= \text{coefficient of } \eta^n \text{ in } \pi^*L(X(n)).
\end{aligned}$$

From eqs. (17) and (18) of §8 we then obtain

$$\begin{aligned}
\mathrm{Sign}(X(n)) &= \text{coefficient of } \eta^n \text{ in } [f_s'(\eta)Q_s(\eta)^{n+1}\times \\
&\qquad \times \sum_{n_0\geq 0} c_{n_0 0\ldots 0}\, Q_s(\eta)^{-n_0}\,\eta^{n_0}\,] \\
&= \text{coefficient of } \eta^n \text{ in } [f_s'(\eta)Q_s(\eta)^{n+1}\times \\
&\qquad \times e^{H(f_s(\eta),0,\ldots,0)}/(1-f_s(\eta)^2)^{e(X)/2}\,].
\end{aligned}$$

The substitution $t=f_s(\eta)$ then permits us to rewrite this as

$$\mathrm{Sign}(X(n)) = \mathrm{res}_{\eta=0}\left[\frac{f_s'(\eta)\,d\eta}{\eta^{n+1}} \frac{e^{H(f_s(\eta),0,\dots,0)}}{(1-f_s(\eta)^2)^{e(X)/2}} Q_s(\eta)^{n+1}\right]$$

$$= \mathrm{res}_{t=0}\left[\frac{dt}{t^{n+1}} \frac{e^{H(t,0,\dots,0)}}{(1-t^2)^{e(X)/2}}\right],$$

or

$$\sum_{n=0}^{\infty} t^n\, \mathrm{Sign}(X(n)) = (1-t^2)^{-e(X)/2}\, e^{H(t,0,\dots,0)}. \tag{2}$$

Finally one obtains from the definition of H (eq. (16) of §8) and the fact that $e_0=1$ is zero-dimensional that

$$H(t,0,\dots,0) = \sum_{c=0}^{s} \langle g_{s+1-c}(t\, e_0)L_c(X),[X]\rangle$$

$$= \langle g_1(t)\,(\mathrm{Sign}(X)\, z),[X]\rangle$$

$$= \mathrm{Sign}(X)\, g_1(t), \tag{3}$$

which, substituted into (2) gives (since $g_1(t) = \tanh^{-1} t = \frac{1}{2}\log\frac{1+t}{1-t}$)

$$\sum_{n=0}^{\infty} t^n\, \mathrm{Sign}(X(n)) = (1-t^2)^{-e(X)/2} \left(\frac{1+t}{1-t}\right)^{\mathrm{Sign}(X)/2}, \tag{4}$$

the formula which was to be proved.

Even simpler is the complete evaluation of $L(X(n))$ when $X = S^{2s}$. Here the basis is just $e_0=f_1=1$, $e_1=f_0=z$ $(b=1)$, and the L-class is of course trivial. Therefore the function $H(t_0,\dots,t_{b-1})$ is just

$$H(t_0) = \sum_{c=0}^{s} \langle g_{s+1-c}(t_0 e_0)L_c(X),[X]\rangle = 0. \tag{5}$$

Of course the Euler characteristic is **two**, so the power series $e^{H(t_0,\dots,t_{b-1})}/(1-t_0^2)^{e(X)/2}$ is just $(1-t_0^2)^{-1}$, and therefore the factor G of the formula for $L(X(n))$ is simply

$$f_s'(\eta) \sum_{n_0=0}^{\infty} Q_s(\eta)^{-2n_0}\, \eta^{2n_0} = \frac{f_s'(\eta)}{1-f_s(\eta)^2} .$$

Therefore the whole expression for $L(X(n))$ is given by

<u>Proposition 1</u>: Let $X = S^{2s}$, <u>n</u> a positive integer, and $\eta \in H^{2s}(X(n))$ the usual element $z_1+\dots+z_n$. Then the L-class of $X(n)$ is given by

$$L(X(n)) = \frac{f_s'(\eta)}{1 - f_s(\eta)^2} \left(\frac{\eta}{f_s(\eta)}\right)^{n+1} . \qquad (6)$$

Notice that this formula automatically has the property that the coefficient of η^n in it is equal to one or to zero according as $\underline{n}$ is even or odd. This is in accordance with (4), since $e(X) = 2$ and $\mathrm{Sign}(X) = 0$ for an even-dimensional sphere. For the case $s=1$, $X(n)$ is the nth symmetric produce of $P_1\mathbb{C}$, and can be naturally identified with $P_n\mathbb{C}$; then $\eta \in H^2(P_n\mathbb{C})$ is the standard generator and (6) reduces (since $f_1(t) = \tanh t$) to the usual relation

$$L(P_n\mathbb{C}) = \left(\frac{\eta}{\tanh \eta}\right)^{n+1} . \qquad (7)$$

We will consider this special case $X = S^2$ more thoroughly when discussing the equivariant version of our theorem.

We now come to the last application of this section. Assume that X is a Riemann surface of genus $\underline{g}$. For the basis $f_0,\ldots,f_{b-1}$ of the cohomology we choose the standard generators of $H^1(X)$, i.e.

$$f_0,\ldots,f_{b-1} = z,\ \alpha_1,\ldots,\alpha_g,\ \alpha_1',\ldots,\ \alpha_g' \qquad (8)$$

with $z \in H^2(X)$ as usual and the α's one-dimensional classes satisfying:

$$\alpha_i\alpha_j = \alpha_i'\alpha_j' = 0 \text{ (all } i,j),\ \alpha_i\alpha_j' = \alpha_i'\alpha_j = 0\ (i \neq j),\ \alpha_i\alpha_i' = -\alpha_i'\alpha_i = z. \qquad (9)$$

Then the dual basis $\{e_i\}$ is $\{1,-\alpha_1',\ldots,-\alpha_g',\alpha_1,\ldots,\alpha_g\}$. To apply the procedure of §8, we must now introduce new variables $t_0,\ldots,t_{b-1}$-- here relabelled $t,t_1,\ldots,t_g,t_1',\ldots,t_g'$ to parallel the labelling of the f_i's-- which are subject to the commutation rules

$$t_it_j = -t_jt_i,\ t_it_j' = -t_j't_i,\ t_i't_j' = -t_j't_i' \quad (1 \leq i,j \leq g). \qquad (10)$$

In particular $t_i^2 = t_i'^2 = 0$. Of course the first variable $\underline{t}$ commutes with everything, since it corresponds to an even-dimensional cohomology class.

Since $L(X) = 1$ and $s=1$ in our case, formula (16) of §8 becomes

$$H = H(t,t_1,\ldots,t_g,t_1',\ldots,t_g') = \langle g_2(t+\delta), [X] \rangle , \qquad (11)$$

where we have used δ to denote the quantity

$$\delta = -\sum_{i=1}^{g} t_i\alpha_i' + \sum_{i=1}^{g} t_i'\alpha_i \quad \epsilon \quad H^1(X)[t_1,\ldots,t_g,t_1',\ldots,t_g']. \tag{12}$$

Because $\underline{t}$ commutes with δ, we can expand the expression $g_2(t+\delta)$ appearing in (11) as a Taylor series in powers of δ. Moreover, δ is a one-dimensional cohomology class and $[X]$ is a two-dimensional homology class, so only the term $g_2''(t)\ \delta^2/2!$ of the Taylor series contributes, so

$$H = \frac{g_2''(t)}{2} < \delta^2, [X] > = -g_2''(t)\ (t_1t_1' + \ldots + t_gt_g'), \tag{13}$$

where to obtain the last line we have used (9), (10) and (12).

We now set

$$s_i = -t_it_i' \quad (1\leq i\leq g) \tag{14}$$

and observe that the various s_i's commute with one another but that

$$s_i^{\,2} = t_it_i't_it_i' = -t_it_it_i't_i' = 0, \tag{15}$$

in view of the fact that $t_i^{\,2}=0$. Therefore we obtain from (13):

$$\begin{aligned} e^H &= e^{g_2''(t)\ (s_1+\ldots+s_g)} \\ &= \prod_{i=1}^{g} e^{s_ig_2''(t)} \\ &= \prod_{i=1}^{g} [\ 1 + s_ig_2''(t) + s_i^{\,2}g_2''(t)^2/2! + \ldots\ ] \\ &= \prod_{i=1}^{g} [\ 1 + s_ig_2''(t)\ ] \\ &= \sum_{r=0}^{g} g_2''(t)^r \sum_{1\leq i_1<\ldots<i_r\leq g} s_{i_1}\ldots s_{i_r}. \end{aligned} \tag{16}$$

We now recall that, since s=1 in our case, we have

$$f_s(\eta) = \tanh\eta, \quad Q_s(\eta) = \eta/\tanh\eta, \quad f_s'(\eta) = \operatorname{sech}^2\eta\ ,$$

and also $e(X) = -2g+2$. Therefore if we substitute (16) into the recipe for calculating G (§8, eqs. (17) and (18)), we obtain

$$\begin{aligned} G = \operatorname{sech}^2\eta\ (1-\tanh^2\eta)^{g-1} \sum_{r=0}^{g} g_2''(\tanh\eta)^r\ (\eta/\tanh\eta)^{-2r}\times \\ \times \sum_{1\leq i_1<\ldots<i_r\leq g} [1(\alpha_{i_1}')1(\alpha_{i_1})\ldots 1(\alpha_{i_r}')1(\alpha_{i_r})]\ . \end{aligned} \tag{17}$$

Furthermore, we deduce from the multiplication laws (9) that

$$[1(\alpha'_{i_1})1(\alpha_{i_1})\dots 1(\alpha'_{i_r})1(\alpha_{i_r})] = [1(\alpha'_{i_1})1(\alpha_{i_1})]\dots[1(\alpha'_{i_r})1(\alpha_{i_r})]. \quad (18)$$

We write

$$\gamma_i = [1(\alpha'_i)1(\alpha_i)] \in H^2(X(\infty)) \qquad (1\le i\le g). \quad (19)$$

Then the substitution into (17) of equation (18) and of the fact that $1 - \tanh^2\eta = \operatorname{sech}^2\eta$ gives

$$G = (\operatorname{sech}\eta)^{2g} \sum_{r=0}^{g} \left(\frac{g_2''(\tanh\eta)}{\eta^2/\tanh^2\eta}\right)^r \sum_{1\le i_1<\dots<i_r\le g} \gamma_{i_1}\dots\gamma_{i_r}$$

$$= (\operatorname{sech}\eta)^{2g} \prod_{i=1}^{g} [\,1 + \gamma_i \frac{\tanh^2\eta}{\eta^2} g_2''(\tanh\eta)\,] \quad (20)$$

However, from the definition of $g_s(t)$, it is clear that its derivative is precisely $g_{s-1}(t)/t$. Moreover, $g_1(t) = \tanh^{-1}t$. Therefore

$$g_2''(t) = \frac{d}{dt}\left(\frac{\tanh^{-1}t}{t}\right) = \frac{1}{t(1-t^2)} - \frac{\tanh^{-1}t}{t^2}. \quad (21)$$

Inserting this in (20) produces

$$G = \prod_{i=1}^{g} [\operatorname{sech}^2\eta + \gamma_i\left(\frac{\tanh\eta - \eta\operatorname{sech}^2\eta}{\eta^2}\right)]. \quad (21)$$

Therefore the L-class of the symmetric product X(n) is given by

$$L(X(n)) = \left(\frac{\eta}{\tanh\eta}\right)^{n+1} G = \left(\frac{\eta}{\tanh\eta}\right)^{n-2g+1} \prod_{i=1}^{g} [\frac{\eta^2}{\sinh^2\eta} +$$

$$\gamma_i\left(\frac{\tanh\eta - \eta\operatorname{sech}^2\eta}{\tanh^2\eta}\right)]. \quad (22)$$

Finally, we can use the following considerations to simplify (22). Write

$$\xi_i = [1(\alpha_i)],\ \xi'_i = [1(\alpha'_i)] \in H^1(X(\infty)),\quad \sigma_i = \xi_i\xi'_i \in H^2(X(\infty)), \quad (23)$$

so that we have, on restricting to H*(X(n)),

$$\sigma_i = \sum_{k=1}^{n}\sum_{l=1}^{n} (\pi_k^*\alpha_i)(\pi_l^*\alpha'_i) = \sum_{k\ne l} (\pi_k^*\alpha_i)(\pi_l^*\alpha'_i) + \sum_{k=1}^{n} \pi_k^*(\alpha_i\alpha'_i)$$

$$= [1(\alpha_i)1(\alpha'_i)] + [1(z)] = -\gamma_i + \eta. \quad (24)$$

But now we can repeat an argument used earlier in this section, namely

$$\sigma_i{}^2 = \xi_i\xi_i'\xi_i\xi_i' = -\xi_i\xi_i\xi_i'\xi_i' = 0$$

since the ξ_i's are one-dimensional, and therefore power series in σ_i break off after the first power. Thus for any power series $\underline{f}$,

$$f(\gamma_i) = f(\eta - \sigma_i) = f(\eta) - \sigma_i f'(\eta).$$

Applying this to the power series $t/\tanh t$ gives

$$\begin{aligned} \frac{\gamma_i}{\tanh \gamma_i} &= \frac{\eta}{\tanh \eta} - \sigma_i\left(\frac{1}{\tanh \eta} - \frac{\eta}{\sinh^2 \eta} \right) \\ &= \frac{\gamma_i}{\tanh \eta} + \frac{\eta(\eta - \gamma_i)}{\sinh^2 \eta} . \end{aligned} \tag{25}$$

Substituting this into eq. (22) gives finally

<u>Theorem 1</u> (Macdonald [25]): Let X be a Riemann surface of genus $\underline{g}$ and $\underline{n}$ a positive integer. Choose a basis for $H^1(X)$ as in (9) and define the classes γ_i and η in $H^2(X(n))$ as above. Then

$$L(X(n)) = \left(\frac{\eta}{\tanh \eta} \right)^{n-2g+1} \prod_{i=1}^{g} \frac{\gamma_i}{\tanh \gamma_i} . \tag{26}$$

What Macdonald in fact proved was more. Namely, X(n) is known to be a complex manifold when X is a Riemann surface, and Macdonald showed that its Chern class is given by

$$c(X(n)) = (1 + \eta)^{n-2g+1} \prod_{i=1}^{g} (1 + \gamma_i). \tag{27}$$

We have introduced the notations α_i, α_i', ξ_i, ξ_i', σ_i, η used above in accordance with the notation of Macdonald's paper.

§14. The equivariant case

If a finite group G acts on X, then the diagonal action of G on X^n induces an action on X(n) sending an unordered n-tuple $\{x_1,\dots,x_n\}$ to the n-tuple $\{g\circ x_1,\dots,g\circ x_n\}$. If X is an even-dimensional manifold as in §8, then X(n) is a rational homology manifold and L(g,X(n)) is defined by the theory given in §4; from the result of §5 we obtain the formula for the calculation of L(g,X(n))

$$\pi^* L(g,X(n)) \quad = \quad \sum_{\sigma\in S_n} L(g\circ\sigma, X^n), \tag{1}$$

where for $\sigma \in S_n$ the automorphism $g\circ\sigma = \sigma\circ g$ of X^n is the map sending $(x_1,\dots,x_n)$ to $(g\circ x_{\sigma(1)},\dots,g\circ x_{\sigma(n)})$. We therefore must calculate $L(g\circ\sigma, X^n)$ in order to evaluate the equivariant L-class on X(n), and this can be done by the Atiyah-Singer theorem. The proof is fairly similar to the proof for L(X(n)) given in the last four sections, and we will therefore give a briefer account of it, only emphasizing the points of difference with the non-equivariant result.

Just as before, if σ is a product of two permutations, then $g\circ\sigma$ also acts on X^n as a product of the corresponding two operations. Therefore, since any permutation is a product of cyclic permutations, and since the equivariant L-class is symmetric, it suffices to consider $L(g\circ\sigma_r, X^r)$, where σ_r is a cyclic permutation of $\{1,\dots,r\}$. Similarly, since the equivariant Euler class e(g,X) defined in §9 (cf. equation (6) of §9) and the signature Sign(g,X) defined in §2 are multiplicative, we can express the values of these invariants for X(n) in terms of the corresponding numbers for the cyclic permutation σ_r. For example, Sign(g,X(n)) is given by a formula exactly corresponding to (12) of §9, with the difference that τ_r now denotes $\mathrm{Sign}(g\circ\sigma_r, X^r)$, and e(g,X(n)) is given by a similar formula with τ_r replaced by $e(g\circ\sigma_r, X^r)$. The values of the equivariant signature and Euler characteristic will be computed below for $g\circ\sigma_r$; we give here the resulting formulas for X(n):

$$\sum_{n=0}^{\infty} t^n\, e(g,X(n)) \quad = \quad \prod_{r=1}^{\infty} \exp\left[\ \frac{t^r}{r} e(X^{g^r})\ \right], \tag{2}$$

$$\sum_{n=0}^{\infty} t^n \operatorname{Sign}(g,X(n)) = \prod_{\substack{r=1 \\ r \text{ even}}}^{\infty} \exp\left[\frac{t^r}{r}\, e(X^{g^r})\right] \prod_{\substack{r=1 \\ r \text{ odd}}}^{\infty} \exp\left[\frac{t^r}{r} \operatorname{Sign}(g^r,X)\right] \tag{3}$$

The first step is to compute the fixed-point set of $g\circ\sigma_r$. This consists of the set of r-tuples $(x_1,\dots,x_r)$ with

$$(x_1,\dots,x_r) = (g\circ x_2,\dots,g\circ x_r,g\circ x_1),$$

from which we conclude that $x_1 = g\circ x_2 = g^2\circ x_3 = \dots = g^r\circ x_1$. Thus

$$(X^r)^{g\circ\sigma_r} = \{(x,g^{-1}\circ x,\dots,g^{1-r}\circ x) \mid g^r x = x\}. \tag{4}$$

In particular, the fixed-point set of $g\circ\sigma_r$ on X^r is isomorphic to the fixed-point set of g^r on X. As a first consequence, we obtain from the Lefschetz formula (eq. (7) of §9) that

$$e(g\circ\sigma_r,X^r) = e((X^r)^{g\circ\sigma_r}) = e(X^{g^r}), \tag{5}$$

and equation (2) follows. If we let $\underline{j}$ denote the inclusion in X of the fixed-point set of g^r, $\underline{i}$ the inclusion in X^r of the fixed-point set of $g\circ\sigma_r$, $d = d_r$ the diagonal map $X \to X^r$, and $\hat{g}$ the map from X^r to itself sending $(x_1,\dots,x_r)$ to $(x_1,g^{-1}\circ x_2,\dots,g^{1-r}\circ x_r)$, and $\underline{h}$ the isomorphism of X^{g^r} onto $(X^r)^{g\circ\sigma_r}$ given by (4), we have from (4) the commutative diagram

$$\begin{array}{ccccc} X^{g^r} & & \xrightarrow[\approx]{h} & & (X^r)^{g\circ\sigma} \\ j\downarrow\subset & & & & \subset\downarrow i \\ X & \xrightarrow[\subset]{d} & X^r & \xrightarrow[\approx]{\hat{g}} & X^r \end{array} \tag{6}$$

This will be used to study the inclusion $\underline{i}$ and in particular the corresponding Gysin homomorphism $i_! = (\hat{g}\circ d\circ j\circ h^{-1})_! = \hat{g}_! d_! j_! h_!^{-1}$. Here the Gysin homomorphisms of $\hat{g}$ and $\underline{h}$ are easy to compute since one has in general $f_! = (f^*)^{-1} = (f^{-1})^*$ for an isomorphism $f:X\to Y$. The Gysin homomorphism of d_r was calculated in §10. Finally, it will turn out from the formula for $L'(g\circ\sigma_r,X^r)$ that the Gysin homomorphism $j_!$ only needs to be applied to $L'(g^r,X) \in H^*(X^{g^r})$, on which it of course gives $L(g^r,X)$.

Leaving till last the calculation of $L'(g\circ\sigma_r,X^r)$, we show

how to deduce from it a formula for $L(g,X(n))$. We proceed as in the non-equivariant case (§11). Thus for $A = (j_1,\dots,j_r)$ with the j_i's distinct elements of $N = \{1,\dots,n\}$, we define a projection map π_A from X^n to X^r by

$$\pi_A(x_1,\dots,x_n) = (x_{j_1},\dots,x_{j_r}) \tag{7}$$

and set

$$L(A) = \pi_A^* L(go\sigma_r,X^r) \in H^*(X^n). \tag{8}$$

The only difference with the non-equivariant situation in this part of the calculation is that $L(A)$ depends on the order of the j_i's rather than just on the unordered set A. To see what this dependence is, set

$$L_r' = h^*L'(go\sigma_r,X^r) \in H^*(X^{g^r}); \tag{9}$$

this is the class which will be calculated below. It depends on $\underline{r}$, $\underline{g}$ and X but not on σ_r (i.e. if we replace σ_r by another cyclic permutation of $\{1,\dots,r\}$, or equivalently renumber the $\underline{r}$ factors of X^r, the value of L_r' is unchanged); this will follow from the evaluation of L_r' below and can also be seen directly. We then set

$$L_r = j_! L_r' \in H^*(X). \tag{10}$$

Then by the remarks following diagram (6) we have

$$L(go\sigma_r,X^r) = \hat{g}^*(d_! L_r) \tag{11}$$

and therefore

$$L(A) = (\hat{g}o\pi_A)^*(d_! L_r). \tag{12}$$

The map $\hat{g}o\pi_A$ from X^n to X^r is

$$\hat{g}o\pi_A(x_1,\dots,x_n) = (x_{j_1},g^{-1}ox_{j_2},\dots,g^{-r+1}ox_{j_r}). \tag{13}$$

Clearly, if $A' = (j_{\tau(1)},\dots,j_{\tau(r)})$ for some $\tau\in S_r$ is an r-tuple with the same elements as A but in a different order, then

$$\hat{g}o\pi_{A'}(x_1,\dots,x_n) = (x_{j_{\tau(1)}},g^{-1}ox_{j_{\tau(2)}},\dots,g^{-r+1}ox_{j_{\tau(r)}})$$

$$= \tau \circ (g^{-\tau^{-1}(1)+1} \circ x_{j_1}, \ldots, g^{-\tau^{-1}(r)+1} \circ x_{j_r})$$

$$= \tau \circ \hat{g}_\tau \circ \pi_A (x_1,\ldots,x_n), \tag{14}$$

where

$$\hat{g}_\tau = g^{-\tau^{-1}(1)+1} \times \ldots \times g^{-\tau^{-1}(r)+1} : X^r \to X^r. \tag{15}$$

Since $d_! L_r \in H^*(X^r)$ is clearly symmetric, i.e. fixed under the action of S_r on $H^*(X^r)$, we have

$$L(A') = (\hat{g} \circ \pi_{A'})^*(d_! L_r) = (\tau \circ \hat{g}_\tau \circ \pi)^*(d_! L_r) = \pi_A^* \hat{g}_\tau^* \tau^*(d_! L_r)$$

$$= \pi_A^* \, \hat{g}_\tau^* \, (d_! L_r). \tag{16}$$

This shows the dependence of $L(A')$ on the order of the elements of A'. It follows that the sum of $L(A')$ over all A' with the same underlying set as A is given by

$$\sum_{A'} L(A') = r! \, \pi_A^*(T(d_! L_r)), \tag{17}$$

where $T: H^*(X^r) \to H^*(X^r)$ is the average of the maps $\hat{g}_\tau^*$:

$$T(e_1 \times \ldots \times e_r) = \frac{1}{r!} \sum_{\tau \in S_r} \hat{g}_\tau^* (e_1 \times \ldots \times e_r)$$

$$= \sum_{\substack{i_1,\ldots,i_r \\ \text{a permutation} \\ \text{of } 0,-1,\ldots,-r+1}} (g^*)^{i_1} e_1 \times \ldots \times (g^*)^{i_r} e_r. \tag{18}$$

In the non-equivariant case, the right-hand side of (17) was simply $r! \, L(A)$. It follows that the term $(r-1)! \, L(A)$ occurring in (6) and (7) of §11, and corresponding to the $(r-1)!$ possible cyclic permutations of a set A of <u>r</u> elements, must in the equivariant situation be replaced by $(r-1)! \pi_A^*(T(d_! L_r))$, and therefore that equation (7) of §11 becomes

$$L(g, X(n)) = \partial_N [e^K], \tag{19}$$

$$K = K(x_1,\ldots,x_n) = K_1 + K_2 + \ldots, \tag{20}$$

where

$$K_r \; = \; \sum_{\substack{A\subset N \\ |A|=r}} (r-1)!\; x_A\; \pi_A^*(T(d_! L_r)). \tag{21}$$

If we now substitute for $d_!$ the formula given in §10, namely

$$d_! L_r \; = \; \{((\pi_1\times 1)^*a)\ldots((\pi_r\times 1)^*a)(1\times L_r)\}/[X], \tag{22}$$

and proceed as in the non-equivariant case (cf. the derivation of eqs. (10), (15) of §11), we find

$$K_r \; = \; \frac{1}{r}\; T(\; \{\bar{a}^r\,(1\times L_r)\}/[X]\;), \tag{23}$$

where $\bar{a}$ is the element defined in §11(14).

We now calculate the classes L_r' and L_r. Set $2m = \dim(X^{g^r})$.

Proposition 1: The class L_r' defined in (9) is given by

$$L_r' \; = \; \begin{cases} e(X^{g^r})\,[X^{g^r}], & \text{if } \underline{r} \text{ is even,} \\ \sum\limits_{c\geq 0} r^{c-m} L_c'(g^r,X), & \text{if } \underline{r} \text{ is odd,} \end{cases} \tag{24}$$

where $[X^{g^r}]$ denotes the fundamental class in cohomology of the fixed-point set of g^r on X (with the relevant twisted coefficients if X^{g^r} is not oriented), $e(X^{g^r})$ is the Euler characteristic, and $L_c'(g^r,X)$ is the component in $H^{2c}(X^{g^r})$ of $L'(g^r,X)\in H^*(X^{g^r})$. Therefore the class $L_r = j_! L_r' \in H^*(X)$ is given by

$$L_r \; = \; \begin{cases} e(X^{g^r})\; z, & \text{if } \underline{r} \text{ is even,} \\ r^{-m}\sum\limits_{c\geq 0} r^c L_{c+s-m}(g^r,X), & \text{if } \underline{r} \text{ is odd,} \end{cases} \tag{25}$$

where $\underline{z}$ is as usual the fundamental class in cohomology of X, and $L_c(g^r,X)$ is the component in $H^{2c}(X)$ of $L(g^r,X)\in H^*(X)$.

Proof: To apply the G-signature formula, we must calculate the normal bundle of X^{g^r} embedded in X^r as the fixed-point set of $g\circ\sigma_r$ (i.e. embedded by the map $\hat{g}\circ d\circ j$ in the notation of (6)), the eigenvalues of the action of $g\circ\sigma_r$ on this bundle over X^{g^r}, the corresponding eigenbundles, and their characteristic classes. The idea is very simple but the details a little complicated because of the special rôle played by the eigenvalue -1 in the Atiyah-Singer formula. To make it more clear what the splitting is, we first consider the case of a complex manifold X with group action. A point $x \in X$ with $g^r \circ x = x$ is identified under the inclusion map $\hat{g}dj$ with the point $(x, g^{-1}\circ x, \ldots, g^{-r+1}\circ x)$, which we denote $\bar{x}$, so

$$T_{\bar{x}}(X^r) = T_x X \oplus T_{g^{-1}\circ x} X \oplus \ldots \oplus T_{g^{-r+1}\circ x} X \quad . \tag{26}$$

We use V to denote the vector space $T_x X$, and identify $T_{g^{-i+1}\circ x} X$ with V by the map induced by g^{i-1}. Then the decomposition (26) becomes

$$T_{\bar{x}}(X^r) = V \oplus \ldots \oplus V \quad (\underline{r} \text{ copies}) \tag{27}$$

with $g\circ\sigma_r$ acting by

$$(v_1, \ldots, v_r) \mapsto (v_2, \ldots, v_r, g^r v_1), \qquad (v_1, \ldots, v_r \in V) \tag{28}$$

where $g^r v_1$ refers to the action of g^r on V which is the differential of the map g^r of X. No action of $\underline{g}$ appears before the vectors $v_2, \ldots, v_r$ because we have already used the map g in identifying (26) and (27); the map g^r then appears because we have used $\underline{g}$ $r-1$ times to identify $T_{g^{-r+1}\circ x} X$ with V and therefore the map $\underline{g}$ from $T_{g^{-r+1}\circ x} X$ to $T_{g^{-r}\circ x} X = V$ gets identified with the map g^r from V to V.

So far everything is the same as in the real case. However, in the complex case we can immediately deduce the eigenvalues of $g\circ\sigma_r$ from (28), namely $(v_1, \ldots, v_r)$ is an eigenvector of $\lambda \in \mathbb{C}$ if

$$(v_2, \ldots, v_r, Av_1) = (\lambda v_1, \ldots, \lambda v_r), \tag{29}$$

where we have used A to denote the linear map $g^r : V \to V$, and clearly this is the same as

$$Av_1 = \lambda v_r = \lambda^2 v_{r-1} = \ldots = \lambda^r v_1 . \tag{30}$$

That is, λ is an eigenvalue of $g\circ\sigma_r$ exactly when λ^r is an eigenvalue of $g^r|V$, and the eigenspaces correspond under

$$v \leftrightarrow (v, \lambda v, \ldots, \lambda^{r-1} v) \qquad (v \in V, \ Av = \lambda^r v). \tag{31}$$

If we remove the eigenspace of eigenvalue 1, which corresponds to passing from the restriction to $(X^r)^{g\circ\sigma_r}$ of $T(X^r)$ to the normal bundle $N^{g\circ\sigma_r}$ of $(X^r)^{g\circ\sigma_r}$, we therefore have found the splitting of this normal bundle (henceforth denoted N) into eigenbundles:

namely, for $\lambda \neq 1$ the complex subbundle N_λ of N on which $g\circ\sigma_r$ acts as multiplication by λ is given by

$$N_\lambda \cong M_{\lambda^r} \tag{32}$$

if $\lambda^r \neq 1$ (here M denotes the normal bundle of X^{g^r} in X, and M_ϑ the complex subbundle of M on which g^r acts as multiplication by ϑ) and by

$$N_\lambda \cong TX^{g^r} \tag{33}$$

if $\lambda^r = 1$. We write L_λ for the characteristic class L_θ of §2, where $\lambda = e^{i\theta}$, i.e. for a complex bundle ξ with Chern class $\Pi(1+x_j)$,

$$L_\lambda(\xi) = \prod_{x_j} \frac{\lambda e^{2x_j}+1}{\lambda e^{2x_j}-1}. \tag{34}$$

Then the Atiyah-Singer formula together with (32), (33) gives

$$h^* L'(g\circ\sigma_r, X^r) = L(X^{g^r}) \prod_{\substack{\lambda^r=1\\ \lambda\neq 1}} L_\lambda(TX^{g^r}) \prod_\vartheta \prod_{\lambda^r=\vartheta} L_\lambda(M_\vartheta). \tag{35}$$

We now use the identity

$$\prod_{\lambda^r=\vartheta} \frac{\lambda z+1}{\lambda z-1} = \begin{cases} 1, & \text{if } \underline{r} \text{ is even,} \\ \dfrac{\vartheta z^r+1}{\vartheta z^r-1}, & \text{if } \underline{r} \text{ is odd.} \end{cases} \tag{36}$$

We find that if $\underline{r}$ is even, (35) reduces to $\prod x_j$, where x_j ranges over the formal roots of the Chern class of TX^{g^r}, and this product is just the Euler class of TX^{g^r}. If $\underline{r}$ is odd, (35) reduces to

$$\prod_{x_j(TX^{g^r})} \left(\frac{x_j}{\tanh x_j} \cdot \frac{\coth rx_j}{\coth x_j}\right) \prod_\vartheta \prod_{x_j(M_\vartheta)} \left(\frac{\vartheta e^{2rx_j}+1}{\vartheta e^{2rx_j}-1}\right), \tag{37}$$

and if we multiply this by r^m (where $m = \dim_{\mathbb{C}} X^{g^r}$) it becomes exactly the Atiyah-Singer expression for $L'(g^r, X)$ except that each two-dimensional class x_j is replaced by rx_j, i.e. it is the even-dimensional cohomology class whose component of degree 2c is r^c times the 2c-dimensional component of $L'(g^r, X)$. This proves (24) in the complex case.

We now turn to the real case, where the idea is similar but the splitting up into eigenspaces more involved. Since the eigenvalues of $g\circ\sigma_r$ on N are the rth roots of the eigenvalues of g^r on M (the notations are the same as those introduced for X complex except that N and M are now real bundles), and since the eigenvalue -1 receives special treatment in the Atiyah-Singer formula, we will have two cases according as -1 is an rth root of +1 or of -1, i.e. according as $\underline{r}$ is even or odd.

For a kxk matrix A, we define an rkxrk matrix $F_r(A)$ by

$$F_r(A) = \begin{vmatrix} 0 & I & 0 & \dots & 0 \\ 0 & 0 & I & \dots & 0 \\ & & \dots & & \\ 0 & 0 & 0 & \dots & I \\ A & 0 & 0 & \dots & 0 \end{vmatrix} , \qquad (38)$$

where I denotes a kxk identity matrix and 0 a kxk zero matrix. Then (28) states that the action of $g\circ\sigma_r$ on $T_{\bar{x}}(X^r)$ in terms of a basis given by the isomorphism (27) is given by the matrix $F_r(\Lambda)$, where Λ is the matrix of the action of g^r on $T_xX=V$. We split up the normal bundle M of X^{g^r} in X as in §2 (eq. (26)), namely as the direct sum of a real bundle N_π of dimension $2s(\pi)$ on which g^r acts as multiplication with -1 and of real bundles N_θ $(0<\theta<\pi)$ of dimension $2s(\theta)$ on which g^r acts with respect to a suitable basis as the matrix A_θ of eq. (24) of §2 (or rather, as the Kronecker product $A_\theta \otimes I_{s(\theta)}$). Then, since $V = T_x(X^{g^r}) \oplus M$, and g^r acts as the identity on the vector space $T_x(X^{g^r})$ of dimension $2m = \dim X^{g^r} = 2s - 2s(\pi) - \sum_{0<\theta<\pi} s(\theta)$, we obtain

$$\Lambda \cong (1)\otimes I_{2m} \oplus (-1)\otimes I_{2s(\pi)} \oplus \sum_{0<\theta<\pi} A_\theta \otimes I_{s(\theta)}, \qquad (39)$$

where (1) and (-1) denote the corresponding 1x1 matrices. We must therefore write $F_r(1)$, $F_r(-1)$, and $F_r(A_\theta)$ as a direct sum of matrices 1, -1, and A_θ in order to apply the Atiyah-Singer formula to $g\circ\sigma_r$. This obviously gives ($\approx$ denotes similarity of matrices)

$$F_r(1) \approx 1 \oplus A_{2\pi/r} \oplus A_{4\pi/r} \oplus \dots \oplus A_{(r-1)\pi/r} \qquad (\underline{r}\text{ odd}), \qquad (40)$$

$$F_r(1) \approx 1 \oplus -1 \oplus A_{2\pi/r} \oplus \dots \oplus A_{(r-2)\pi/r} \qquad (\underline{r}\text{ even}), \qquad (41)$$

$$F_r(-1) \approx -1 \oplus A_{\pi/r} \oplus A_{3\pi/r} + \dots \oplus A_{(r-2)\pi/r} \qquad (\underline{r}\text{ odd}), \qquad (42)$$

$$F_r(-1) \approx A_{\pi/r} \oplus \dots \oplus A_{(r-1)\pi/r} \qquad (\underline{r}\text{ even}), \qquad (43)$$

$$F_r(A_\theta) \approx A_{\theta/r} \oplus A_{(\theta+2\pi)/r} \oplus \dots \oplus A_{(\theta+(r-1)\pi)/r} \quad (\text{all } \underline{r}). \tag{44}$$

From equations (39)-(44) we obtain the eigenvalue decomposition of the normal bundle N. Namely, if $\underline{r}$ is <u>even</u>, then

$$N = N_\pi \oplus \sum_{k=1}^{\frac{r}{2}-1} N_{2\pi k/r} \oplus \sum_{\substack{k=1\\ k \text{ odd}}}^{r-1} N_{\pi k/r} \oplus \sum_{\substack{0<\theta<\pi\\ r\varphi\equiv\theta \bmod 2\pi}} N_\varphi , \tag{45}$$

where

$$N_\pi \cong T(X^{g^r}), \tag{46}$$

$$N_{2\pi k/r} \cong T(X^{g^r})\otimes\mathbb{C} \quad (\text{for } k=1,2,\dots,\tfrac{r}{2}-1), \tag{47}$$

$$N_{\pi k/r} \cong M_\pi\otimes\mathbb{C} \quad (\text{for } k=1,3,\dots,r-1), \tag{48}$$

$$N_\varphi \cong M_\theta \quad (\text{for } 0<\theta<\pi,\ \ r\varphi\equiv\theta \pmod{2\pi}). \tag{49}$$

If $\underline{r}$ is <u>odd</u>, then

$$N = \sum_{k=1}^{\frac{r-1}{2}} N_{2\pi k/r} \oplus N_\pi \oplus \sum_{\substack{k=1\\ k \text{ odd}}}^{r-2} N_{\pi k/r} \oplus \sum_{\substack{0<\theta<\pi\\ r\varphi\equiv\theta \bmod 2\pi}} N_\varphi , \tag{50}$$

where

$$N_{2\pi k/r} \cong T(X^{g^r})\otimes\mathbb{C} \quad (\text{for } k=1,2,\dots,\tfrac{r-1}{2}), \tag{51}$$

$$N_\pi \cong M_\pi , \tag{52}$$

$$N_{\pi k/r} \cong M_\pi\otimes\mathbb{C} \quad (\text{for } k=1,3,\dots,r-2), \tag{53}$$

$$N_\varphi \cong M_\theta \quad (\text{for } 0<\theta<\pi,\ \ r\varphi\equiv\theta \pmod{2\pi}). \tag{54}$$

We denote the Chern classes of the complex bundles $T(X^{g^r})\otimes\mathbb{C}$, $M_\pi\otimes\mathbb{C}$, and M_θ by $\Pi_{j=1}^{2m}(1+x_j)$, $\Pi_{j=1}^{2s(\pi)}(1+x_j^\pi)$, and $\Pi_{j=1}^{s(\theta)}(1+x_j^\theta)$, respectively. Then substituting (45)-(49) in the G-signature formula (eq. (27) of §2) gives, for $\underline{r}$ even,

$$h^*L'(g\circ\sigma_r, X^r) = L(X^{g^r})\cdot L_\pi(T(X^{g^r}))\cdot \prod_{k=1}^{\frac{r}{2}-1} L_{\frac{2\pi k}{r}}(T(X^{g^r})\otimes\mathbb{C}) \cdot \prod_{\substack{k=1\\ k \text{ odd}}}^{r-1} L_{\frac{\pi k}{r}}(M_\pi\otimes\mathbb{C}) \cdot \prod_{\substack{0<\theta<\pi\\ r\varphi\equiv\theta \bmod 2\pi}} L_\varphi(M_\theta). \tag{55}$$

Now $L_\pi(\xi) = e(\xi)L(\xi)^{-1}$ by definition for a real bundle ξ. If we substitute this in (55) we see that all the L-classes cancel (this follows either from (36) with $\underline{r}$ even or directly, if we note that $\coth(\theta+i\pi/2) = \tanh\theta = 1/\coth\theta$, and therefore $L_\theta(\xi)L_{\theta+\pi/2}(\xi) = 1$ for any complex bundle ξ, so that the L-classes occurring in (55) cancel in pairs), and therefore

$$h^*L'(g\circ\sigma_r, X^r) \;=\; e(T(X^{g^r})) \;=\; e(X^{g^r})\;[X^{g^r}], \quad (\underline{r}\text{ even}) \tag{56}$$

where the first $\underline{e}$ denotes the Euler class, the second $\underline{e}$ the Euler characteristic, and the square brackets the fundamental class in cohomology. If, on the other hand, $\underline{r}$ is $\underline{\text{odd}}$, we get from (50)-(54):

$$\begin{aligned} h^*L'(g\circ\sigma_r, X^r) \;=\; & L(X^{g^r})\cdot \prod_{k=1}^{(r-1)/2} L_{\frac{2\pi k}{r}}(T(X^{g^r})\otimes\mathbb{C})\cdot L_\pi(M_\pi) \\ & \cdot \prod_{\substack{k=1\\ k\text{ odd}}}^{r-2} L_{\frac{\pi k}{r}}(M_\pi\otimes\mathbb{C}) \cdot \prod_{0<\theta<\pi}\;\prod_{r\varphi\equiv\theta \bmod 2\pi} L_\varphi(M_\theta) \end{aligned} \tag{57}$$

$$\begin{aligned} = \; & \prod_{j=1}^{2m}\left(\frac{x_j}{\tanh x_j}\prod_{k=1}^{(r-1)/2}\coth\left(x_j+\frac{i\pi k}{r}\right)\right)\cdot e(M_\pi)\,\cdot \\ & \prod_{j=1}^{2s(\pi)}\left(\frac{\tanh x_j^\pi}{x_j^\pi}\prod_{\substack{k=1\\ k\text{ odd}}}^{r-2}\coth\left(x_j^\pi+\frac{i\pi k}{2r}\right)\right)\cdot \\ & \prod_{0<\theta<\pi}\prod_{j=1}^{s(\theta)}\left(\prod_{r\varphi\equiv\theta \bmod 2\pi}\coth\left(x_j^\theta+\frac{i\varphi}{2}\right)\right) \end{aligned}$$

$$= e(M_\pi)\prod_{j=1}^{2m}\frac{x_j}{\tanh rx_j}\;\prod_{j=1}^{2s(\pi)}\frac{\tanh rx_j^\pi}{x_j^\pi}\;\prod_{0<\theta<\pi}\prod_{j=1}^{s(\theta)}\coth rx_j^\theta, \tag{58}$$

where the last line has been obtained by using the identity (36) for $\underline{r}$ odd. Clearly, if we multiply expression (58) by r^m, then it is exactly the Atiyah-Singer expression for $L'(g^r, X^r)$ except that each two-dimensional class x_j has been multiplied by $\underline{r}$. This proves the case $\underline{r}$ odd of (24), and therefore together with eq. (56) proves Proposition 1.

We now return to the calculation of $L(g, X(n))$. Because L_r is homogeneous of top dimension for $\underline{r}$ even (eq. (25)), we can compute

$d_! L_r$ directly, without using the result of §10 on the Gysin homomorphism of the diagonal map $\underline{d}$. Therefore for even $\underline{r}$ we can replace (22) by

$$d_! L_r \;=\; e(X^{g^r})\, d_! z \;=\; e(X^{g^r})\, z\times\ldots\times z, \qquad (\underline{r}\text{ even}). \tag{59}$$

Since the action of G on X preserves the orientation, $T(z\times\ldots\times z) = z\times\ldots\times z$ (here T is the map (18)). Therefore (21) reduces to

$$K_r \;=\; e(X^{g^r}) \sum_{1\leq j_1<\ldots<j_r\leq n} (r-1)!\; x_{j_1}\ldots x_{j_r}\, (\pi^*_{j_1} z)\ldots(\pi^*_{j_r} z)$$

$$=\; \frac{e(X^{g^r})}{r}\, (x_1\pi^*_1 z + \ldots + x_n\pi^*_n z)^r, \qquad (\underline{r}\text{ even}), \tag{60}$$

or, in the notation of §12, simply

$$K_r \;=\; \frac{e(X^{g^r})}{r}\, t_0^r, \qquad (\underline{r}\text{ even}). \tag{61}$$

From (23) and (25), on the other hand, we get

$$K_r \;=\; \Sigma_{c=0}^{m}\, r^{c-m-1}\, T(\,\{\mathbf{z}^r(1\times L_{c+s-m}(g^r,X))\}/[X]\,), \qquad (\underline{r}\text{ odd}), \tag{62}$$

where $m = (\dim X^{g^r})/2$ depends on $\underline{r}$. Formulas (61) and (62), combined with (19) and (20) and the definitions of ${}_N$, t_i, and so on as given in §§11-12, provide a complete evaluation of the class $L(g,X(n))$.

Just as in §12, we want to study the dependence of $L(g,X(n))$ on $\underline{n}$, and we know from our experience there that, to do so, we must study the dependence of $K = K(t_0,\ldots,t_b)$ and of $G = e^K = G(t_0,\ldots,t_b)$ on t_b. By eq. (61), $K_r = K_r(t_0)$ for even $\underline{r}$, so the dependence of K on t_b only comes from the odd $\underline{r}$ terms (62). Just as in §12 (cf. eqs. (9), (10)), we can write out $\mathbf{z}$ as $\Sigma\, t_i\times e_i$ and note that, since $e_b = z$ has the same degree as $[X]$, the only dependence of (62) on t_b must come from monomials of the form $(t_0\times e_0)^{r-1}(t_b\times e_b)L_0(g^r,X)$. Since $c\geq 0$ and $s\geq m$, we have $c+s-m>0$ unless $c=0$ and $s=m$. Since X is connected, X^{g^r} can only have the same dimension as X if it equals X, i.e. if $g^r = \mathrm{id}$. This is a further difference from the non-equivariant case, where the L-class always had a non-zero leading coefficient. If $g^r = \mathrm{id}$, then $s=m$, $L(g^r,X) = L(X)$, and $L_0(X) = 1$. We have therefore proved that

$$K(t_0,\ldots,t_b) \;=\; K(t_0,\ldots,t_{b-1},0) + t_b \sum_{\substack{r=1\\ r\text{ odd}\\ g^r=\mathrm{id}}}^{\infty} r^{-s}\, t_0^{r-1}. \tag{63}$$

If g has even order, then g^r is never the identity for odd r, so we find from (63) that K and hence also G are independent of t_b. This does not, however, mean that $L(g,X(n))$ has especially nice stability properties. To the contrary, $L(g,X(n))$ and $L(g,X(n+1))$ are completely unrelated for r odd. For if we expand

$$G \;=\; \Sigma\, c_{n_0\ldots n_{b-1}}\; t_0^{n_0}\ldots t_{b-1}^{n_{b-1}} \tag{64}$$

(the sum is taken over all $n_0,\ldots,n_{b-1}\geqslant 0$), we find

$$\partial_N G \;=\; \Sigma\, c_{n_0\ldots n_{b-1}}\; [n_0(f_0)\ldots n_{b-1}(f_{b-1})], \tag{65}$$

where the sum is taken over all $n_0,\ldots,n_{b-1}$ with $n_0+\ldots+n_{b-1} = n$, and therefore the values of $\partial_N G$ for n and for $n+1$ involve entirely different coefficients of the expansion (64).

If, however, the element g has odd order p (we assume that G acts effectively, so that g has the same order as an element of G and as an automorphism of X), then we obtain from (63) that

$$G = G(t_0,\ldots,t_b) \;=\; e^{t_b\,\varphi(t_0)}\; G(t_0,\ldots,t_{b-1},0), \tag{66}$$

where $\varphi(t) = p^{-s}t^{p-1} + \ldots$ is the power series

$$\varphi(t) \;=\; \sum_{\substack{k=1\\ k \text{ odd}}}^{\infty} (kp)^{-s}\, t^{kp-1} \;=\; p^{-s}\,\frac{g_s(t^p)}{t}, \tag{67}$$

where $g_s(\;)$ is the power series defined in §8. We now proceed just as in the proof of Proposition 1 of §12. The proof is identical up to (20) of §12, which in our notation states that $\partial_N G$ equals

$$\sum_{j=0}^{n}\sum_{k=0}^{n}\Bigl(\sum_{\substack{B\subset N\\ |B|=k}} z_B\Bigr)\Bigl(\sum_{\substack{A\subset N\\ |A|=j}} \partial_A G(t_0,\ldots,t_{b-1},0)\Bigr)\;\frac{d^k}{dt^k}\bigl(\varphi(t)^{n-j-k}\bigr)\Big|_{t=0}. \tag{68}$$

The first factor equals $\eta^k/k!$, and the second factor is a well-defined element G_j in the stable homology group $H^*(X(\infty))$, namely if we expand $G(t_0,\ldots,t_{b-1},0)$ as in (64) then G_j is given by expression (65) with the sum over all $n_0,\ldots,n_{b-1}$ satisfying $n_0+\ldots+n_{b-1} = j$ (a finite sum). The last factor can be evaluated by setting $y = g_s(t^p)$:

$$\frac{d^k}{dt^k}\bigl(\varphi(t)^{n-j-k}\bigr) \;=\; k!\cdot \text{coefficient of } t^k \text{ in } \Bigl(p^{-s}\,\frac{g_s(t^p)}{t}\Bigr)^{n-j-k}$$

$$= k! \cdot \text{coefficient of } t^{n-j} \text{ in } (p^{-s}\, g_s(t^p))^{n-j-k} .$$

Clearly this is zero if n-j is not divisible by p, while if n-j=ap it is

$$= k! \operatorname{res}_{x=0}\left[\frac{dx}{x^{a+1}} \;(p^{-s}\, g_s(a))^{ap-k}\right]$$

$$= k! \operatorname{res}_{y=0}\left[\frac{f_s'(y)dy}{f_s(y)^{a+1}} \;(p^{-s}\, y)^{ap-k}\right]$$

$$= k!\, p^{-asp}\, p^{sk} \cdot \text{coefficient of } y^k \text{ in } \frac{y^{ap+1} f_s'(y)}{f_s(y)^{a+1}} .$$

Therefore (68) becomes

$$\partial_N G = \sum_{a=0}^{[n/p]} G_{n-ap} \frac{p^s\, \eta^{ap+1}\, f_s'(p^s\eta)}{f_s(p^s\eta)^{a+1}} . \qquad (69)$$

We have proved the following theorem:

Theorem 1: Let all notations and definitions be as in Theorem 1 of §8. Assume further that g is an orientation-preserving diffeomorphism $X \to X$ of order p. For $j \geq 0$, define $G_j \in H^*(X(\infty))$ by

$$G_j = \sum_{n_0+\dots+n_{b-1}=j} d_{n_0\dots n_{b-1}} [n_0(f_0)\dots n_{b-1}(f_{b-1})], \qquad (70)$$

where the numbers $d_{n_0\dots n_{b-1}} \in \mathbb{Q}$ are the coefficients in the expansion

$$\sum_{n_0,\dots,n_{b-1}\geq 0} d_{n_0\dots n_{b-1}} t_0^{n_0}\dots t_{b-1}^{n_{b-1}}$$

$$= \exp\left[\sum_{\substack{r=2\\ r \text{ even}}}^{\infty} \frac{e(X^{g^r})}{r} t_0^r + \sum_{\substack{r=1\\ r \text{ odd}}}^{\infty} \sum_{c=0}^{s} r^{c-s-1} \langle T\alpha^r \cdot L_c(g^r,X),[X]\rangle\right]. \qquad (71)$$

Here $e(X^{g^r})$ is the Euler characteristic of the fixed-point set of g^r, $L_c(g^r,X)$ is the component in $H^{2c}(X)$ of the equivariant L-class of g^r, $\langle -,[X]\rangle$ is the map from A to B defined in Theorem 1 of §8, and

$$T\alpha^r = \frac{1}{r!} \sum (g^{*i_1}\alpha)\dots(g^{*i_r}\alpha), \qquad \text{(summation as in (18))}, \qquad (72)$$

where $\alpha \in A$ is the element of §8(14) and g^* acts on A via its action on $H^*(X)$. Then if p is even, we have

$$L(g,X(n)) = j^* G_n , \qquad (73)$$

where j^* is the restriction map $H^*(X(\infty)) \to H^*(X(n))$. If p is odd,

$$L(g,X(n)) \;=\; \frac{p^s\eta}{f_s(p^s\eta)}\, f_s'(p^s\eta) \sum_{a=0}^{[n/p]} G_{n-ap}\, Q_{s,p}(\eta)^a, \tag{74}$$

where $Q_{s,p}(\cdot)$ is the power series

$$Q_{s,p}(t) \;=\; \frac{t^p}{f_s(p^s t)} \;=\; p^{-s}\, t^{p-1}\, Q_s(p^s t). \tag{75}$$

Corollary: Let X, g be as in the theorem, $\dim X = 2s$, g of odd order p. Let j denote the inclusion of $X(n)$ in $X(n+p)$. Then

$$j^* L(g,X(n+p)) \;=\; Q_{s,p}(\eta_n)\, L(g,X(n)). \tag{76}$$

It should perhaps be observed that the occurrence here of the inclusion $X(n) \subset X(n+p)$ rather than $X(n) \subset X(n+1)$ is perfectly natural, since it is this map which arises in the equivariant setting. Recall that the map $j: X(n) \subset X(n+1)$ was defined by

$$\{x_1,\ldots,x_n\} \;\to\; \{x_0,x_1,\ldots,x_n\}, \tag{77}$$

with $x_0 \in X$ a fixed basepoint. This map is equivariant only if x_0 is fixed by G. In general, the set X^G is empty and the only way to get an equivariant inclusion map is to use the map

$$\{x_1,\ldots,x_n\} \;\to\; \{x_0, gx_0,\ldots,g^{p-1}x_0, x_1,\ldots,x_n\} \tag{78}$$

from $X(n)$ to $X(n+p)$. That is, instead of mapping an unordered set of n points of X to its union with a fixed point of X, we map it to its union with a whole orbit $Gx_0 \subset X$. This also means that, if we wanted to talk about the action of G on $X(\infty)$, we really should form p different limit spaces $\varinjlim X(n)$, with n ranging over a fixed residue class (mod p) and inclusion maps defined by (78). These spaces would all be homeomorphic, but would not necessarily be the same when considered as G-spaces. However, for our purposes (eq. (74)) we need only the cohomology of $X(\infty)$, and so do not need to enter into these subtleties.

§15. Equivariant L-classes for symmetric products of spheres

In §13 we evaluated the expression previously obtained for $L(X(n))$ in two cases of especially simple nature, namely for X a sphere and for X a Riemann surface. The case of a sphere was much the simpler, and is the only one we are also able to cope with in the equivariant case.

Let X be a sphere S^{2s} on which a finite group G acts orientably. Since $H^*(X)$ consists only of the elements 1 and z, both preserved by G, the group G acts trivially on the cohomology of X and of its products and symmetric products. In particular, the averaging operator T of §14, eq. (18) is the identity, and (using (7) of §9) we also have

$$e(X^{g^r}) = e(g^r,X) = e(X) = 2. \tag{1}$$

Since only $H^0(X)$ and $H^{2s}(X)$ are non-zero, only $L_0(g^r,X)$ and $L_c(g^r,X)$ could be non-zero, and the latter equals $\mathrm{Sign}(g^r,X)$ and is therefore certainly zero since X has a vanishing middle homology group. The class $L_0(g^r,X)$ is one if $g^r = \mathrm{id}$, and zero if $g^r \neq \mathrm{id}$. Therefore the right-hand side of (71) of §14 is

$$\exp\left[\sum_{\substack{r=2\\ r\ \text{even}}}^{\infty} \frac{2}{r}\, t_0^r + \sum_{\substack{r=1\\ r\ \text{odd}\\ g^r=\mathrm{id}}}^{\infty} r^{-s-1} < \alpha^r, [X] > \right] = \frac{1}{1-t_0^2}, \tag{2}$$

since the sum over odd r vanishes ($\alpha = t_0 e_0$, and since $e_0=1$ has degree zero, $<\alpha^r,[X]> = 0$) and the sum over even r equals $-\log(1-t_0^2)$. Therefore the coefficients d_{n_0} of Theorem 1 of §14 (here b=1) are equal to 1 if n_0 is even and to 0 if n_0 is odd. Substituting this into (70) of §14 and using §8(4), we find

$$G_j = \begin{cases} 0, & \text{if } j \text{ is odd,} \\ \eta^j, & \text{if } j \text{ is even.} \end{cases} \tag{3}$$

We then obtain from (73)-(75) of §14 that:

$$L(g,X(n)) = \begin{cases} 0, & \text{if } p \text{ is even, } n \text{ is odd,} \\ \eta^n, & \text{if } p \text{ is even, } n \text{ is even,} \end{cases} \tag{4}$$

while for odd p

$$L(g,X(n)) = \frac{p^s\eta}{f_s(p^s\eta)} f_s'(p^s\eta) \sum_{\substack{0\leq a\leq n/p \\ n\equiv a \bmod 2}} \frac{\eta^n}{f_s(p^s\eta)^a} . \tag{5}$$

Write $q = [n/p]$. The sum in (5) over $0\leq a\leq q$ can be replaced by a sum over $-\infty < a \leq q$, since for negative $\underline{a}$ we have $\eta^n f_s(p^s\eta)^{-a} = 0$ (because the power series f_s begins with a term of positive degree, and $\eta^{n+1} = 0$ in $H^*(X(n))$). Thus it is a sum over $a = q, q-2, q-4, \ldots$ if $q \equiv n \pmod 2$ and over $a = q-1, q-3, \ldots$ if $q \not\equiv n \pmod 2$. Since

$$y^{-q} + y^{2-1} + y^{4-q} + \ldots = \frac{y^{-q}}{1-y^2}$$

and similarly with $\underline{q}$ replaced by $\underline{q}-1$, we obtain from (5) that

$$L(g,X(n)) = \frac{p^s\eta}{f_s(p^s\eta)} \cdot \frac{f_s'(p^s\eta)}{1 - f_s(p^s\eta)^2} \cdot \frac{\eta^n}{f_s(p^s\eta)^{[n/p]}} ,$$

$$(\underline{p} \text{ odd}, \quad n \equiv [n/p] \pmod 2), \tag{6}$$

$$= \frac{p^s\eta}{f_s(p^s\eta)} \cdot \frac{f_s'(p^s\eta)}{1 - f_s(p^s\eta)^2} \cdot \frac{\eta^n}{f_s(p^s\eta)^{[n/p]-1}} ,$$

$$(\underline{p} \text{ odd}, \quad n \not\equiv [n/p] \pmod 2). \tag{7}$$

We state these results as a theorem.

<u>Theorem 1</u>: Let $X = S^{2s}$ be an even-dimensional sphere, $\underline{g}$ a diffeomorphism of X to itself of order $\underline{p}$, preserving the orientation. Then the equivariant L-class of the induced action of $\underline{g}$ on the nth symmetric product $X(n)$ is given by equation (4), (6), or (7), depending on the values of $\underline{p}$, $\underline{n}$, and $n - p[n/p]$ modulo 2. For example, if $X = S^2$ and $\underline{p}$ is odd, then

$$L(g,S^2(n)) = \frac{p\eta^{n+1}}{(\tanh p\eta)^k} , \qquad (\underline{p} \text{ odd}) \tag{8}$$

where

$$k = \begin{cases} [n/p] + 1, & \text{if } n \equiv [n/p] \pmod 2, \\ [n/p], & \text{if } n \not\equiv [n/p] \pmod 2. \end{cases} \tag{9}$$

The last assertion follows since for $s = 1$ we have $f_s(t) = \tanh t$, and therefore $f_s'(t) = 1 - f_s(t)^2 = \operatorname{sech}^2 t$. The case $X = S^2$ is of interest since it provides a verification of the theorem of §14; namely, the nth symmetric product of S^2 is equal to complex projective space $P_n\mathbb{C}$, for which certain equivariant L-classes were computed in §6. To identify $S^2(n)$ with $P_n\mathbb{C}$, we write S^2 as $P_1\mathbb{C}$, so that a point of S^2 is written $(z:w)$, where $(z,w) \in \mathbb{C}^2 - \{0\}$ and $(z:w) = (tz:tw)$ for $t \in \mathbb{C}^*$ $= \mathbb{C} - \{0\}$. For $0 \leq i \leq n$, we define

$$h_i: \quad (\mathbb{C}^2 - \{0\})^n \to \mathbb{C} \tag{10}$$

by

$$h_i((z_1,w_1),\ldots,(z_n,w_n)) = \sum_{\substack{I\subset N\\ |I|=i}} (\prod_{j\in I} z_j)(\prod_{j\notin I} w_j) \tag{11}$$

Clearly, if we multiply (z_j,w_j) by $t\in\mathbb{C}^*$, then each h_i is also multiplied by $\underline{t}$ (since the jth coordinate appears in exactly one of the products in each summand in (11)). Therefore the point $(h_0:\ldots:h_n)$ is unchanged if we replace (z_j,w_j) by (tz_j,tw_j), so we have an induced map

$$\bar{h}: \quad (S^2)^n \to P_n\mathbb{C}, \tag{12}$$

mapping $(z_1:w_1),\ldots,(z_n:w_n)$ to $(h_0:\ldots:h_n)$, where h_i is given by (11) for any representatives (z_j,w_j) of the points in S^2. This map is clearly symmetric, so induces a map

$$h: \quad S^2(n) \to P_n\mathbb{C}. \tag{13}$$

The map sending $(h_0:\ldots:h_n) \in P_n\mathbb{C}$ to the unordered n-tuple of roots (z_j,w_j) of the homogeneous equation $\Sigma_{i=0}^{n} h_i w^i z^{n-i} = 0$ (which is clearly independent of the choice of h's representing the given point of $P_n\mathbb{C}$) is then an inverse to the map h, which therefore is an isomorphism.

Now let $G = \mu_p \subset S^1$ be the group of pth roots of unity, acting on S^2 by

$$\zeta\circ(z:w) = (\zeta z:w) \qquad (\zeta^p = 1,\ (z:w)\in S^2). \tag{14}$$

Then G acts diagonally on $S^2(n)$, sending $\{(z_j:w_j)\}_{j=1,\ldots,n}$ to $\{\zeta\circ(z_j:w_j)\}_{j=1,\ldots,n}$. Using the map $\underline{h}$ to identify $P_n\mathbb{C}$ with $S^2(n)$, we find from (14) and (11) that the induced action on $P_n\mathbb{C}$ is

$$\zeta\circ(h_0:\ldots:h_n) = (h_0:\zeta h_1:\zeta^2 h_2:\ldots:\zeta^n h_n). \tag{15}$$

In particular, the action of G on $P_n\mathbb{C}$ is linear, and for linear actions we calculated the equivariant L-class in §6. Substituting the definition (15) of the action of G into the result for $L(g,P_n\mathbb{C})$ given in eq. (13) of §6, we obtain

$$L(\zeta,S^2(n)) = \sum_{\lambda\in S^1} \prod_{k=0}^{n}\left(\eta \, \frac{\lambda^{-1}\zeta^k e^{2\eta}+1}{\lambda^{-1}\zeta^k e^{2\eta}-1}\right), \tag{16}$$

where $\eta \in H^2(X(n))$ is the usual element, and corresponds under the isomorphism $\underline{h}$ to the Hopf class $x \in H^2(P_n\mathbb{C})$.

We wish to show that (16) agrees with the result of Theorem 1, namely that $L(\zeta,S^2(n))$ is given by (4) or by (8) according as $\underline{p}$ is even or odd. Recall from §6 that (16) is a finite sum, the only non-zero summands being those with λ equal to one of the eigenvalues ζ^k. In particular, λ can only contribute to (16) if it is a pth root of unity. Since we want the element ζ to have order exactly $\underline{p}$, ζ is a primitive p^{th} root of unity and therefore λ must be a power of ζ. Therefore (16) can be rewritten

$$L(\zeta,S^2(n)) = \sum_{j=1}^{n} \prod_{k=0}^{n} \left(\eta \, \frac{\zeta^{k-j} e^{2\eta}+1}{\zeta^{k-j} e^{2\eta}-1} \right), \tag{17}$$

or equivalently, since we might as well assume $\zeta = e^{2\pi i/p}$,

$$L(\zeta,S^2(n)) = \sum_{j=1}^{n} \prod_{k=0}^{n} \left(\eta \, \coth(\eta + \frac{k-j}{p}\pi)\right). \tag{18}$$

It is clear from either of these expressions that

$$L(\zeta,S^2(n+p)) = Q \, L(\zeta,S^2(n)), \tag{19}$$

where

$$Q = \prod_{j=1}^{p} \left(\eta \, \frac{\zeta^{j} e^{2\eta}+1}{\zeta^{j} e^{2\eta}-1} \right). \tag{20}$$

This was evaluated in eq. (36) of §14 :

$$Q = \begin{cases} \eta^p & \text{if } \underline{p} \text{ is even.} \\ \eta^p/\tanh p\,, & \text{if } \underline{p} \text{ is odd.} \end{cases} \tag{21}$$

If we compare this with expression (4) for even $\underline{p}$ or (8) for odd $\underline{p}$, we see that the value of $L(\zeta,S^2(n))$ computed in Theorem 1 also satisfies equation (19). Therefore if the value computed there agrees with (18) for some $\underline{n}$, it also agrees for n+p, and so we only have to check values of $\underline{n}$ which are smaller than $\underline{p}$. If $0 \leq n < p$, the number $\underline{k}$ defined in (9) equals zero or one according as $\underline{n}$ is odd or even, so (8) states

$$L(\zeta,S^2(n)) = \begin{cases} p\,\eta^{n+1}, & \text{if } \underline{n} \text{ is odd} \\ p\eta^{n+1}/\tanh p\eta, & \text{if } \underline{n} \text{ is even} \end{cases} \quad (0 \leq n < p), \tag{22}$$

for odd $\underline{p}$. But $\eta^{n+1} = 0$ in $H^*(P_n\mathbb{C})$, so (22) also agrees with the expression (4) proved for even $\underline{p}$. We therefore only need to prove that (17) and (22) are equal for $0 \leq n < p$. We state this as a lemma.

__Lemma 1__: Let $\underline{n}$, $\underline{p}$ be integers, $0 \leq n < p$, and ζ a primitive p^{th} root of unity. Then the following identity holds:

$$\frac{1}{p} \sum_{\lambda^p=1} \prod_{k=0}^{n} \frac{\lambda\zeta^k z + 1}{\lambda\zeta^k z - 1} = \begin{cases} 1, & \text{if } \underline{n} \text{ is odd,} \\ \dfrac{z^p+1}{z^p-1}, & \text{if } \underline{n} \text{ is even.} \end{cases} \tag{23}$$

__Proof__: Define a rational function $F(z)$ by

$$F(z) = \prod_{k=0}^{n} \frac{\zeta^k z + 1}{\zeta^k z - 1}. \tag{24}$$

Then the left-hand side of (23) is the rational function

$$G(z) = \frac{1}{p} \sum_{\lambda^p=1} F(\lambda z). \tag{25}$$

Since $\underline{n}$ is smaller than $\underline{p}$, the factors in (24) have their poles in distinct points (namely at $z = \zeta^{-k}$, $k=0,1,\dots,n<p$), and therefore $F(z)$ is a rational function which has at most a simple pole for $\underline{z}$ a pth root of unity and is holomorphic everywhere else (including $z=\infty$). Therefore $G(z)$ has the same properties. Moreover, $G(z)$ is invariant under $z \mapsto \theta z$ for any p^{th} root of unity θ, so $G(z)$ is a meromorphic function of z^p, with a simple pole at $z^p=1$ and regular everywhere else. Therefore $G(z)$ must be of the form $a + b/(z^p-1)$ for some constants $\underline{a}$ and $\underline{b}$. To evaluate these, we observe that $F(\infty) = 1$ and $F(0) = (-1)^n$, and therefore $G(z)$ also equals 1 at $z=\infty$ and 0 at $z=0$. It follows that $G(z)$ is the function appearing on the right of (23).

As a corollary of the lemma, the residue of $G(z)$ at $z=1$ equals $2/p$ or 0 according as $\underline{n}$ is even or odd. Since the residue of $F(\lambda z)$ at $z=1$ is clearly 0 for $\lambda \neq \zeta^{-j}$ ($j=0,1,\dots,n$) and $2\,\Pi_{j\neq k}\,(\zeta^{k-j}+1)/(\zeta^{k-j}-1)$ for $\lambda=\zeta^{-j}$, we obtain the identity (independent of $\underline{p}$)

$$\Sigma_{j=0}^{n}\ \Pi_{k=0,\ k\neq j}^{n}[(\zeta^k+\zeta^j)/(\zeta^k-\zeta^j)] = \frac{1+(-1)^n}{2}. \tag{26}$$

CHAPTER III: THE G-SIGNATURE THEOREM AND SOME ELEMENTARY NUMBER THEORY

In this chapter we study the number-theoretical properties of certain elementary trigonometric sums occurring in connection with the G-signature theorem. It is clear from the form of the G-signature theorem that, if the group G is finite, the expression for the equivariant signature $\mathrm{Sign}(g,X)$ involves the evaluation of certain finite sums whose terms are products of the cotangents of rational multiples of π. The motivation for the further study of such cotangent sums arose from two discoveries. One was that the formula given by Brieskorn for the signature of the variety

$$V_a = \{(z_1,\ldots,z_n)\in \mathbb{C}^n \mid \sum_i z_i^{a_i} = 1\}, \qquad a = (a_1,\ldots,a_n)\in \mathbb{Z}_+^n, \tag{1}$$

can be expressed in terms of such a sum. The second was that cotangent sums of this sort appear in the classical literature, and indeed in a variety of contexts: the theory of modular functions, the Hardy-Ramanujan-Rademacher formula for the partition function, the theory of quadratic residues, the theory of indefinite binary quadratic forms, the problem of the class numbers of quadratic fields over $\mathbb{Q}$, and the problem of generating random numbers. We shall say nothing about these classical appearances of cotangent sums (references, however, have been given for all of them), nor--except for a brief remark about the Legendre-Jacobi symbol and the law of quadratic reciprocity--about their connection with the theory of group actions on manifolds. Indeed, it still seems to be mysterious that the same expressions occur in the theory of the transformation of the Dedekind modular function $\eta(z)$ under the action of $SL(2,\mathbb{Z})$ and in the theory of four-dimensional manifolds with group action. The connection between the signature theorem on 4-manifolds and the theory of quadratic extensions of $\mathbb{Q}$, on the other hand, has been accounted for since its discovery by the work of Hirzebruch on the Hilbert modular group and the resolution of certain two-dimensional complex singularities; it manifests itself, for example, in the equality of two invariants associated to a T^2-bundle over S^1, one defined topologically and the other purely number-theoretic.

To see the sort of topological expression which arises, we specialize the G-signature theorem to the case of a component of a fixed-point set X^g that reduces to a single point $\underline{x}$. Thus $\underline{x}$ is an isolated fixed point of $\underline{g}$ and the action of $\underline{g}$ on T_xX is given by eigenvalues $\lambda_j = e^{2i\theta_j} \neq 1$ $(j=1,\ldots,n,\ 2n=\dim X)$, and the G-signature theorem gives

$$\frac{\lambda_1+1}{\lambda_1-1}\cdots\frac{\lambda_n+1}{\lambda_n-1} = i^{-n}\cot\theta_1\ldots\cot\theta_n \tag{2}$$

as the contribution to Sign(g,X) from the component $\{x\}$ of X^g. If $\underline{g}$ has order $\underline{p}$, then θ_j must be a multiple a_j of π/p, where a_j is an integer defined modulo $\underline{p}$; if further the point $\underline{x}$ is an isolated fixed point of all the powers of $\underline{g}$ (except of course of $g^p = 1$), the integers a_j $(j=1,\ldots,n)$ must all be mutually prime to $\underline{p}$. Then the contribution of $\{x\}$ to the sum $\Sigma_{h\in G}$ Sign(h,X) (where G is the cyclic group generated by $\underline{g}$) is equal to

$$\operatorname{def}_x = \operatorname{def}(p;a_1,\ldots,a_n) = \sum_{\substack{\lambda^p=1\\ \lambda\neq 1}} \frac{\lambda^{a_1}+1}{\lambda^{a_1}-1}\cdots\frac{\lambda^{a_n}-1}{\lambda^{a_n}-1}$$

$$= \sum_{k=1}^{p-1}\frac{\zeta^{ka_1}+1}{\zeta^{ka_1}-1}\cdots\frac{\zeta^{ka_n}+1}{\zeta^{ka_n}-1} = i^{-n}\sum_{k=1}^{p-1}\cot\frac{\pi k a_1}{p}\ldots\cot\frac{\pi k a_n}{p}. \tag{3}$$

$(\zeta = e^{2\pi i/p})$

The reason for the interest in the expression Σ Sign(h,X) is its appearance in the formula for the signature of the quotient X/G (§3 (1)). For any submanifold Y of X, we define def_Y (the "signature defect of Y"; see Hirzebruch [17]) as the sum, taken over all $g\in G$ for which Y is a component of X^g, of the Atiyah-Singer expression for the contribution from Y to Sign(g,X). This is of course only non-zero for finitely many manifolds Y, which are necessarily connected and of even codimension in X. Then

$$|G|\ \operatorname{Sign}(X/G) = \sum_{g\in G}\operatorname{Sign}(g,X) = \operatorname{Sign} X + \sum_{Y\neq X}\operatorname{def}_Y, \tag{4}$$

so the numbers def_Y can be thought of as defects specifying the amount by which the formula Sign X = |G| Sign X/G (which holds for free actions) fails to be true. In the special case that G is cyclic and $\underline{x}$ an isolated fixed-point of every $g\in G-\{1\}$, we obtain (3).

Notice that the expression $\operatorname{def}(p;q_1,\ldots,q_n)$ defined in (3) is a rational number. Indeed, in the second to the last formula of (3), we could have taken any primitive p^{th} root of unity for ζ without changing the sum, so the sum belongs to the subfield of the cyclotomic field of order p of elements equal to all their conjugates, and this subfield is precisely $\mathbb{Q}$. Moreover, if $c = \frac{\lambda+1}{\lambda-1}$ where $\lambda^p = 1$, then $(c+1)^p = (c-1)^p$, so c satisfies an algebraic equation with integer coefficients and leading coefficient p, and therefore pc is an algebraic integer. It follows that the expression (3) when multiplied by p^n is a rational integer:

$$p^n \operatorname{def}(p;a_1,\ldots,a_n) \in \mathbb{Z} . \tag{5}$$

We will show in §16 that the p^n in (5) can be replaced by p^{n-1}, or by μ_n, where μ_n is a known integer and is independent of p (it is the denominator of the Hirzebruch L_n-polynomial: see [12]). The same sort of reasoning used to prove (5) shows that def_Y is always a rational number, and indeed an element of $\mathbb{Z}[\frac{1}{d}]$, where $d = |G|$.

These remarks give some idea of the subject matter of this chapter and of the type of interplay which takes place between the topological and the number-theoretical aspects of expressions such as $\operatorname{def}(p;a_1,\ldots,a_n)$. A more precise description of the contents of the chapter is as follows: The purely number-theoretic aspects of the cotangent sums are considered in §16. We first give an elementary treatment of the expression (3) for $n=2$ (this is the sum that appears most in the literature, in connection with modular functions and quadratic fields), showing its connection with the Legendre-Jacobi symbol and giving an elementary proof of a reciprocity law, due to Rademacher, from which the law of quadratic reciprocity can be deduced. We then consider the general case, giving a rational expression for $\operatorname{def}(p;a_1,\ldots,a_n)$ and proving a generalization of the Rademacher reciprocity law; the latter is then used to prove the above-mentioned result on the denominator of $\operatorname{def}(p;a_1,\ldots,a_n)$. The formula giving the signature of the Brieskorn variety as a cotangent sum is also proved. In §17 we construct two explicit group actions for which the statement of the G-signature theorem reduces to the Rademacher reciprocity law (or rather its generalization to higher n). One of these is precisely the action of a product of cyclic groups on $P_n\mathbb{C}$ (the Bott action) which was studied in §6. The other is a direct generalization of the construction given by Hirzebruch [17] for the case n=2, but is included because it requires the theorem of §3 for the L-class of a quotient space and thus provides a further link with the results of Chapter I. In the final section, we give a computation of $\mathrm{Sign}(g,V_a)$, where V_a is the manifold (1) and $\underline{g}$ acts by multiplying z_k with an $a_k{}^{th}$ root of unity. The computation is exactly parallel to Brieskorn's in the case g= id. The result can be rewritten as a trigonometric sum by using the results of §16, the case g=id being the formula for $\mathrm{Sign}(V_a)$ mentioned at the beginning of the introduction. That this formula involves cotangents suggests that it can be obtained by the use of the G-signature theorem, and indeed this can be done: one studies a certain hypersurface in $P_n\mathbb{C}$ invariant under the Bott action. However, this alternate proof will not be given here; it was given by Hirzebruch in the course of a series of lectures in which a direct proof was given of the result of Bott proved in §6 [21]. In one special case, we do give an evaluation using the G-signature theorem. Namely, when the action of G on V_a is free, we can calculate $\mathrm{Sign}(g,V_a)$ by replacing the non-compact manifold V_a with the bounded manifold $V_a \cap D^{2n}$, and if we then glue onto the boundary the D^2-bundle associated to the S^1-bundle $(V_a \cap S^{2n-1}) \to (V_a \cap S^{2n-1})/S^1$, we obtain a closed G-manifold to which the G-signature theorem can be applied.

§16. Elementary properties of cotangent sums

We wish to study the trigonometrical sums

$$\operatorname{def}(p;a_1,\dots,a_n) = (-1)^{n/2} \sum_{k=1}^{p-1} \cot\frac{\pi k a_1}{p} \dots \cot\frac{\pi k a_n}{p} \tag{1}$$

defined in the introduction, where p is a positive integer and the a_i's are integers prime to p. We can write $(-1)^{n/2}$ for i^{-n} in (1) since the sum is clearly zero for odd n (the substitution $k \to p-k$ replaces each cotangent by its negative).

We begin by an elementary treatment of $\operatorname{def}(p;q,r)$. We shall later give a rational expression for (1) which simplifies, when n=2, to

$$\operatorname{def}(p;q,r) = 4p \sum_{k=1}^{p} \left(\left(\frac{kq}{p}\right)\right)\left(\left(\frac{kr}{p}\right)\right), \tag{2}$$

where $((x))$ is the standard notation

$$((x)) = \begin{cases} x - [x] - \frac{1}{2}, & \text{if } x \text{ is not an integer,} \\ 0, & \text{if } x \text{ is an integer.} \end{cases} \tag{3}$$

We can always take r=1 in (2), since r is prime to p and therefore kr runs over all residues (mod p) as k does. Then (2) can be rewritten

$$\operatorname{def}(p;q,1) = -\frac{2}{3}\,(q,p)_D\,, \tag{4}$$

where

$$(q,p)_D = -6p \sum_{k=1}^{p} \left(\left(\frac{k}{p}\right)\right)\left(\left(\frac{kq}{p}\right)\right). \tag{5}$$

This last quantity is the "Dedekind symbol," studied by Dedekind [5] in connection with the behaviour of his modular function $\eta(z)$ under the action of modular transformations (he used the notation (q,p), but we will reserve this notation for its usual meaning as the greatest common divisor of two integers q and p). The reason for the factor 6p in (5) is that it is exactly the factor required to make $(q,p)_D$ an integer.

To study the expression (5), we introduce a slightly more convenient integer-valued function, namely

$$f(p,q) = \sum_{k=1}^{p-1} k\,[\frac{kq}{p}]\,. \tag{6}$$

We also note the formula (valid if $(p,q)=1$, as will always be assumed)

$$\sum_{k=1}^{p-1} [\frac{kq}{p}] = \frac{(p-1)(q-1)}{2}\,. \tag{7}$$

This can be obtained by counting the lattice points below the diagonal of the rectangle $0<x<p$, $0<y<q$, or by substituting $p-k$ for $\underline{k}$ in (6). Since $(p,q)=1$, the set of numbers $qk-p[\frac{qk}{p}]$ runs over a complete set of non-zero residues (mod p) as $\underline{k}$ does, so

$$\sum_{k=1}^{p-1} (qk - p[\frac{qk}{p}])^2 = \sum_{k=1}^{p-1} k^2\,,$$

or, evaluating the various terms,

$$(q^2-1)(p-1)(2p-1) - 12\,q\,f(p,q) + 6\,p \sum_{k=1}^{p-1} [\frac{qk}{p}]^2 = 0. \tag{8}$$

On the other hand,

$$f(q,p) = \sum_{x=1}^{q-1} x\,[\frac{xp}{q}] = \sum_{0<x<q} \Big(\sum_{0<k\leq xp/q} x \Big)$$

$$= \sum_{0<k<p} \Big(\sum_{[\frac{qk}{p}]+1\leq x\leq q-1} x \Big)$$

$$= \sum_{k=1}^{p-1} \frac{1}{2}\Big(q^2 - q - [\frac{qk}{p}]^2 - [\frac{qk}{p}] \Big)$$

$$= \frac{(2q-1)(q-1)(p-1)}{4} - \frac{1}{2}\sum_{k=1}^{p-1} [\frac{qk}{p}]^2\,. \tag{9}$$

Combining (8) and (9), we obtain

$$q\,f(p,q) + p\,f(q,p) = \frac{1}{12}(p-1)(q-1)(8pq-p-q-1). \tag{10}$$

On the other hand, it is easy to write $f(p,q)$ in terms of $(q,p)_D$:

$$f(p,q) = \frac{1}{12}(p-1)(4pq-2q-3p) - \frac{1}{6}(q,p)_D \tag{10A}$$

(notice that it follows immediately from this formula that $(q,p)_D$ is an integer). Therefore we can write (10) in terms of $(q,p)_D$:

$$p\,(p,q)_D + q\,(q,p)_D = \frac{p^2+q^2+1-3pq}{2}\,. \tag{11}$$

Equation (11) is the reciprocity law of Dedekind. Together with the fact that $(q,p)_D$ only depends on the residue class of q mod p, it suffices to define the symbol $(q,p)_D$ entirely (by a Euclidean algorithm). Stated in terms of the defect, eq. (11) becomes

$$\frac{1}{p}\,\mathrm{def}(p;q,1) \;+\; \frac{1}{q}\,\mathrm{def}(q;p,1) \;=\; 1 - \frac{p^2+q^2+1}{3pq}\,. \tag{12}$$

This is the form in which the reciprocity law was generalized by Rademacher [36], who proved that

$$\frac{1}{p}\,\mathrm{def}(p;q,r) + \frac{1}{q}\,\mathrm{def}(q;p,r) + \frac{1}{r}\,\mathrm{def}(r;p,q) = 1 - \frac{p^2+q^2+r^2}{3pqr}\,, \tag{13}$$

where p,q,r are positive and mutually prime. We will prove later a generalization of (13) for the higher defect symbols defined in (1).

First, however, we complete the "elementary" part of this section by relating the Dedekind symbol $(q,p)_D$ and the Legendre-Jacobi symbol $(\frac{q}{p})$. This symbol is defined whenever q,p are relatively prime integers with p odd (e.g. as the sign of the permutation on $\mathbb{Z}/p\mathbb{Z}$ induced by multiplication with q), and is given, just as in the special case of prime p, by Gauß's lemma, namely

$$\left(\frac{q}{p}\right) = (-1)^{N_{q,p}}, \qquad (p \text{ odd},\ (q,p)=1), \tag{14}$$

where

$$N_{q,p} = \left|\left\{ x:\ 1 \le x \le \frac{p-1}{2},\quad qx - p\left[\frac{qx}{p}\right] > \frac{p}{2} \right\}\right|. \tag{15}$$

Therefore, modulo 2, we have

$$N_{q,p} \equiv \sum_{\substack{0<x<p/2\\ [2qx/p]\ \text{odd}}} 1 \equiv \sum_{x=1}^{(p-1)/2} \left[\frac{2qx}{p}\right]$$

$$\equiv \sum_{\substack{k=1\\ k\ \text{even}}}^{p-1} \left[\frac{kq}{p}\right] \equiv \sum_{k=1}^{p-1} (k-1)\left[\frac{kq}{p}\right] = f(p,q) - \frac{(p-1)(q-1)}{2}\,. \tag{16}$$

From (14) we deduce

$$\left(\frac{q}{p}\right) \equiv 2N_{q,p} + 1 \pmod 4, \tag{17}$$

and combining equations (10A) (16), and (17), we obtain the desired relation* between the Dedekind and Legendre-Jacobi symbols:

* This was known to Dedekind (Crelle <u>83</u> (1877) 262-292; Gesammelte Werke, Band I, 174-201, §6).

$$\left(\frac{q}{p}\right) + (q,p)_D \equiv \frac{p+1}{2} \pmod 4. \tag{18}$$

Since $(\frac{q}{p})$ can only take on the values ± 1, it is completely determined by (18), so that a knowledge of the Dedekind symbol also gives the Legendre-Jacobi symbol. Also, if we substitute (18) into the Dedekind reciprocity law (11), we obtain, after a short calculation,

$$\left(\frac{q}{p}\right) + \left(\frac{p}{q}\right) \equiv \frac{(p-1)(q-1)}{2} + 2 \pmod 4, \qquad p,q \text{ odd}, \ (p,q)=1,$$

which is precisely the law of quadratic reciprocity.

We now turn to the general case of definition (1). Our first goal is a rational expression for $\mathrm{def}(p;a_1,\ldots,a_n)$ generalizing equation (2). This is given by

Theorem 1: Let p be a positive integer, and a_j $(j=1,\ldots,n)$ integers prime to p. Then

$$\mathrm{def}(p;a_1,\ldots,a_n) = 2^n p \sum_{\substack{1\le k_1,\ldots,k_n\le p \\ p\,|\,a_1k_1+\ldots+a_nk_n}} \left(\left(\frac{k_1}{p}\right)\right)\ldots\left(\left(\frac{k_n}{p}\right)\right). \tag{19}$$

Proof: We will make frequent use of the well-known dual formulas

$$\sum_{\lambda^p=1} \lambda^r = \begin{cases} 0, & \text{if } p\nmid r, \\ p, & \text{if } p\,|\,r, \end{cases} \tag{20}$$

$$\sum_{k=1}^{p} \lambda^k = \begin{cases} 0, & \text{if } \lambda \neq 1, \\ p, & \text{if } \lambda = 1, \end{cases} \tag{21}$$

where in the latter formula λ is any p^{th} root of unity. Our proof of eq. (19) will be based on the following formula of Eisenstein:

$$\left(\left(\frac{a}{p}\right)\right) = \frac{1}{2p} \sum_{\substack{\lambda^p=1 \\ \lambda\neq 1}} \lambda^{-a} \frac{\lambda+1}{\lambda-1} = \frac{-1}{2p} \sum_{k=1}^{p-1} \sin\frac{2\pi ka}{p} \cot\frac{\pi k}{p}. \tag{22}$$

This can be proved in several ways; the easiest is to note that the difference of the right-hand side for $a=b$ and $a=b-1$ is (using (20))

$$\frac{1}{2p} \sum_{\lambda} \lambda^{-b} \frac{\lambda+1}{\lambda-1} (1-\lambda) = \frac{-1}{2p} \sum_{\lambda} (\lambda^{1-b} + \lambda^{-b})$$

$$= \begin{cases} \frac{2-p}{2p}, & \text{if } b \equiv 0,1 \pmod p \\ 1/p & \text{if } b \not\equiv 0,1 \pmod p \end{cases} = \left(\left(\frac{b}{p}\right)\right) - \left(\left(\frac{b-1}{p}\right)\right).$$

From (20) we obtain the formula

$$\sum_{\substack{\lambda_1^p=1\\ \lambda_1\neq 1}} \frac{\lambda_1+1}{\lambda_1-1}\left(\frac{1}{p}\sum_{k=1}^{p}\lambda_1^{-k}\lambda^{ak}\right) = \begin{cases} \dfrac{\lambda^a+1}{\lambda^a-1}, & \text{if } \lambda\neq 1,\\ 0, & \text{if } \lambda=1,\end{cases}$$

where λ is any p^{th} root of unity. Substituting this into (1) gives

$$\operatorname{def}(p;a_1,\dots,a_n) = \sum_{\substack{\lambda^p=1\\ \lambda\neq 1}} \frac{\lambda^{a_1}+1}{\lambda^{a_1}-1}\cdots\frac{\lambda^{a_n}+1}{\lambda^{a_n}-1}$$

$$= \sum_{\lambda^p=1}\prod_{j=1}^{n}\left[\sum_{\lambda_j^p=1,\ \lambda_j\neq 1}\frac{\lambda_j+1}{\lambda_j-1}\left(\frac{1}{p}\sum_{k_j=1}^{p}\lambda_j^{-k_j}\lambda^{+a_jk_j}\right)\right]$$

$$= \frac{1}{p^n}\sum_{\substack{\lambda^p=\lambda_1^p=\dots=\lambda_n^p=1\\ \lambda_1,\dots,\lambda_n\neq 1}}\ \sum_{1\leq k_1,\dots,k_n\leq p} \frac{\lambda_1+1}{\lambda_1-1}\cdots\frac{\lambda_n+1}{\lambda_n-1}\times$$

$$\times\ \lambda^{+a_1k_1+\dots+a_nk_n}\ \lambda_1^{-k_1}\dots\lambda_n^{-k_n}.$$

The sum over λ can now be evaluated using (20), and the sum over λ_j $(j=1,\dots,n)$ using (22). The result is precisely (19).

Since the denominator of $((\frac{k}{p}))$ is at most $2p$, it follows from the theorem that

$$p^{n-1}\operatorname{def}(p;a_1,\dots,a_n)\ \in\ \mathbb{Z}\,. \tag{23}$$

This is a sharpening of eq. (5) of the introduction, and will be further improved later.

The second theorem we state for the numbers $\operatorname{def}(p;a_1,\dots,a_n)$ is a generalization of the Rademacher reciprocity law (12).

Theorem 2: Let $a_0,\dots,a_{2k}$ be positive and mutually prime integers. Then

$$\sum_{j=0}^{2k}\frac{1}{a_j}\operatorname{def}(a_j;a_0,\dots,\hat{a}_j,\dots,a_{2k}) = 1-\frac{L_k(p_1,\dots,p_k)}{a_0\cdots a_{2k}}\,. \tag{24}$$

Here $\hat{a}_j$ denotes the omission of a_j, and $L_k(p_1,\dots,p_k)$ is the kth Hirzebruch L-polynomial in the variables $p_j=\sigma_j(a_0^2,\dots,a_{2k}^2)$, where σ_j is the jth elementary symmetric polynomial.

<u>Proof</u>: Consider the poles of the rational function

$$f(z) = \frac{1}{2z} \prod_{j=0}^{2k} \frac{z^{a_j} + 1}{z^{a_j} - 1} . \tag{25}$$

These clearly lie at $0, \infty$, and those points $\underline{z}$ on the unit circle with $z^{a_j} = 1$ for some $\underline{j}$. Since the a_j's are prime to one another, it can only happen that $z^{a_i} = z^{a_j} = 1$ if $\underline{z}$ itself is 1; therefore the pole of $\underline{f}$ at a point $\underline{z}$ with $z^{a_j}=1$, $z\neq1$ is simple and its residue can be calculated immediately from (25). Applying the residue theorem gives

$$0 = \sum_{t\in\mathbb{C}} \operatorname{res}_{z=t} (f(z) \, dz)$$

$$= \operatorname{res}_{z=0} + \operatorname{res}_{z=\infty} + \sum_{j=0}^{2k} \sum_{t^{a_j}=1}^{t\neq1} \operatorname{res}_{z=t} + \operatorname{res}_{z=1}$$

$$= \frac{-1}{2} + \frac{-1}{2} + \sum_{j=0}^{2k} \frac{1}{a_j} \prod_{\substack{i=0 \\ i\neq j}}^{2k} \frac{t^{a_j} + 1}{t^{a_j} - 1} + \operatorname{res}_{z=1}(f(z)\, dz).$$

Finally, we find on substituting e^{2t} for $\underline{z}$ that

$$\operatorname{res}_{z=1}(f(z) \, dz) = \operatorname{res}_{t=0}(f(e^{2t}) \cdot 2e^t \, dt)$$

$$= \operatorname{res}_{t=0} (\coth a_0 t \ldots \coth a_{2k} t \, dt)$$

$$= \frac{1}{a_0 \cdots a_{2k}} \operatorname{res}_{t=0} \left[\prod_{j=0}^{2k} \frac{a_j t}{\tanh a_j t} \frac{dt}{t^{2k+1}} \right] .$$

By definition of the L-polynomials, however,

$$\prod_{j=0}^{2k} \frac{a_j t}{\tanh a_j t} = \sum_{r=0}^{\infty} L_r(p_1,\ldots,p_r) \, t^{2r}. \tag{26}$$

Therefore

$$\operatorname{res}_{z=1} (f(z) \, dz) = \frac{L_k(p_1,\ldots,p_k)}{a_0 \cdots a_{2k}} ,$$

and the proof of the theorem is complete.

The case k=1 of Theorem 2 is just eq. (12). For k=2, the right-hand side of (24) is

$$1 - (7p_2 - p_1^2)/45 \text{ abcde,}$$

where a,b,c,d,e are positive and mutually prime integers and

$$p_1 = a^2 + b^2 + c^2 + d^2 + e^2, \quad p_2 = a^2b^2 + \ldots + d^2e^2.$$

Just as the numbers def(p;q,r) turned out to be multiples of 1/3 (eq. (4)), we can deduce from the reciprocity law for k=2 that 45 def(a;b,c,d,e) is an integer. In general, let μ_k be the denominator of L_k, i.e. the smallest positive integer such that $\mu_k L_k(p_1,\ldots,p_k)$ is a polynomial with integer coefficients. Thus $\mu_1=3$, $\mu_2=45$, and in general (see [11])

$$\mu_k = \prod l^{\left[\frac{2k}{l-1}\right]}, \tag{27}$$

where l runs over all odd primes. If we multiply equation (24) by $\mu_k\, a_0\ldots a_{2k}$, the right-hand side is certainly an integer. Because the a_j's are relatively prime, and because $\mathrm{def}(a_j;a_0,\ldots,\hat{a}_j,\ldots,a_{2k})$ is in $\mathbb{Z}[\frac{1}{a_j}]$ (by eq. (23)), we can deduce from this that

$$\mu_k\, \mathrm{def}(a_j;a_0,\ldots,\hat{a}_j,\ldots,a_{2k}) \in \mathbb{Z}. \tag{28}$$

This is only true by the above argument if all of the a_j's are prime to one another. However, to see that $\mu_k \mathrm{def}(p;a_1,\ldots,a_n)$ is always an integer (here n=2k), we observe that its value only depends on the residue classes of a_i (mod p), and that a_i is prime to p; therefore we can use Dirichlet's theorem to replace each a_j by a large prime without changing its class in $\mathbb{Z}/p\mathbb{Z}$, and so we can assume that the a_j's are prime to one another as well as to p. Combining this with (23) and the formula (27) for μ_k, we obtain:

Theorem 3: Let $p\geq 1$ and $a_1,\ldots,a_n$ (n=2k) be integers with $(a_j,p)=1$ for all j. Then $\mathrm{def}(p;a_1,\ldots,a_n)$ is a rational number whose denominator divides μ_k (independently of p and the a_i's). More precisely, the denominator is at most equal to

$$\prod l^{\left[\frac{n}{l-1}\right]}, \tag{29}$$

where the product runs over the odd prime divisors l of p.

We note that Theorem 3 is quite sharp. If p is prime to μ_k, of course, it states that $\mathrm{def}(p;a_1,\ldots,a_n)$ is an integer (this is the case if p has no prime factor $\leq n+1$), but if p and μ_k do have a common factor, then the bound for the denominator in (29) really can occur.

For n=2, for instance, def(p;q,r) is an integer if and only if $\underline{p}$ is not a multiple of 3 (this can be seen easily from the treatment of $(q,p)_D$ given at the beginning of the section). For higher $\underline{n}$, we can test the sharpness of (29) by taking $\underline{p}$ small. Thus

$$\operatorname{def}(3;1,\ldots,1) = \sum_{j=1}^{2}\left(\cot\frac{j\pi}{3}\right)^{2k} = \left(\frac{1}{\sqrt{3}}\right)^{2k} + \left(\frac{-1}{\sqrt{3}}\right)^{2k} = \frac{2}{3^k}$$

has denominator exactly 3^k, and

$$\operatorname{def}(5;1,\ldots,1) = \sum_{j=1}^{4}\left(\cot\frac{j\pi}{5}\right)^{2k}$$

$$= \left(\sqrt{\frac{5+2\sqrt{5}}{5}}\right)^{2k} + \left(\sqrt{\frac{5-2\sqrt{5}}{5}}\right)^{2k} + \left(-\sqrt{\frac{5-2\sqrt{5}}{5}}\right)^{2k} + \left(-\sqrt{\frac{5+2\sqrt{5}}{5}}\right)^{2k}$$

$$= \frac{2}{5^k}\left[\,(5+2\sqrt{5})^k + (5-2\sqrt{5})^k\,\right]$$

has denominator at most $5^{[k/2]}$ (exactly $5^{[k/2]}$ unless $\underline{k}$ is an odd multiple of 5).

The last trigonometric sum evaluated in this section will be the one giving the signature of the Brieskorn manifold V_a (eq. (1) of the introduction to the chapter). It differs from the previously considered sums in that the expression given for it will be an integer rather than just a rational number. The reason is that the sum only involves cotangents of the form $\cot\frac{j\pi}{2p}$ ($\underline{j}$ odd). Just as we observed that $c = \frac{\lambda+1}{\lambda-1}$ is 1/p times an algebraic integer if λ is a pth root of unity (cf. the discussion leading to (3) of the introduction), we see that $\underline{c}$ itself is an algebraic integer if $\lambda^p = -1$, since then the equation $(c+1)^p + (c-1)^p = 0$ leads to an algebraic equation for $\underline{c}$ with leading coefficient 1. Thus $\cot\frac{j\pi}{2p}$ ($\underline{j}$ odd) is an algebraic integer, and a rational expression involving such cotangents is therefore a rational integer. We now state

<u>Theorem 4</u>: Let $a_1,\ldots,a_n \geq 2$ be integers, and

$$t(a_1,\ldots,a_n) = \left|\left\{(x_1,\ldots,x_n) \in \mathbb{Z}^n : 0 < x_j < a_j \ (j=1,\ldots,n),\ 0 < \frac{x_1}{a_1} + \ldots + \frac{x_n}{a_n} < 1 \pmod 2\right\}\right|$$

$$- \left|\left\{x \in \mathbb{Z}^n : 0 < x_j < a_j,\ 1 < \frac{x_1}{a_1} + \ldots + \frac{x_n}{a_n} < 2 \pmod 2\right\}\right| \qquad (30)$$

be the expression given by Brieskorn [3] for the signature of the variety V_a. Here $0<y<1 \pmod 2$ means that $2k<y<2k+1$ for some integer k. Then $t(a) = t(a_1,\ldots,a_n)$ is zero if $\underline{n}$ is even, and

$$t(a) = \frac{(-1)^{\frac{n-1}{2}}}{N} \sum_{\substack{j=1 \\ j \text{ odd}}}^{2N-1} \cot\frac{j\pi}{2N}\cot\frac{j\pi}{2a_1}\cdots\cot\frac{j\pi}{2a_n} \tag{31}$$

if $\underline{n}$ is odd, where N is any multiple of the integers $a_1,\ldots,a_n$.

Proof: The statement for even $\underline{n}$ follows immediately on replacing each x_j by $a_j - x_j$ in (30). We could prove (31) by one of the methods used earlier in the section (e.g. by a residue or Fourier series technique or using the formula of Eisenstein), but prefer the following more elementary approach. Let $N>0$ be as in the theorem and set $b_j = N/a_j$ $(j=1,\ldots,n)$. Define a polynomial $f(t)$ by

$$f(t) = \sum_{\substack{0<x_1<a_1 \\ \vdots \\ 0<x_n<a_n}} t^{b_1x_1+\ldots+b_nx_n} \tag{32}$$

$$= \prod_{j=1}^{n} \left(t^{b_j} + t^{2b_j} + \ldots + t^{(a_j-1)b_j} \right)$$

$$= \prod_{j=1}^{n} \frac{t^{b_j} - t^N}{1 - t^{b_j}} . \tag{33}$$

If we write c_r for the coefficient of t^r in $f(t)$, it is clear that the number $t(a_1,\ldots,a_n)$ is precisely $c_1+\ldots+c_{N-1}-c_{N+1}-\ldots-c_{2N-1}+c_{2N+1}+\ldots+c_{3N-1}-\ldots$, i.e.

$$t(a) = \operatorname{res}_{t=0}\left[f(t^{-1})g(t)\frac{dt}{t}\right] \tag{34}$$

where

$$g(t) = t+\ldots+t^{N-1} - t^{N+1} - \ldots - t^{2N-1} + t^{2N+1} + \ldots$$

$$= (t+\ldots+t^{N-1})/(1+t^N) = \frac{t-t^N}{1-t}\cdot\frac{1}{1+t^N} . \tag{35}$$

The residue in (34) is well-defined since $f(t)$ is a polynomial and $f(t^{-1})$ is therefore meromorphic at $t=0$. Since $f(t)$ is a polynomial, moreover, $f(t^{-1})$ has no other pole, and it is clear from (35) that $g(t)$ has poles only at points $\underline{t}$ with $t^N = -1$, these being simple. The residue theorem therefore gives

$$t(a) = - \sum_{z^N = -1} f(z) \operatorname{res}_{t=z}\left[\frac{g(t)dt}{t}\right]$$

$$= - \sum_{t^N = -1} \prod_{j=1}^{n} \frac{t^{b_j} + 1}{1 - t^{b_j}} \cdot \frac{t+1}{1-t} \cdot \frac{-1}{N} .$$

This is equivalent to equation (31).

Notice that the theorem gives even more information about the cotangent sum appearing than the remarks preceding the theorem, for from these remarks it only follows that a cotangent sum involving terms $\cot \frac{j\pi}{2a}$ must be an integer, while it follows from (31) that the cotangent sum appearing there is actually a multiple of N.

It is possible to prove a large number of similar results. For example, a specialization of a very slight generalization of Theorem 4 (see Hirzebruch [18]) yields an integer expression for the quantity

$$\frac{(-1)^{n/2}}{p} \sum_{\substack{j=1 \\ j \text{ odd}}}^{2p-1} \cot \frac{ja_1\pi}{2p} \cdots \cot \frac{ja_n\pi}{2p}$$

$$= \frac{1}{p} \left[\operatorname{def}(2p; a_1, \ldots, a_n) - \operatorname{def}(p; a_1, \ldots, a_n) \right] ,$$

where p is a positive integer and the a_i's are odd integers prime to p. This expression gives a formula for the Browder-Livesay of the free involution T on the lens space $\mathcal{L}(p; a_1, \ldots, a_n)$ defined as the covering translation of the double covering

$$\mathcal{L}(p; a_1, \ldots, a_n) \longrightarrow \mathcal{L}(2p; a_1, \ldots, a_n).$$

However, we have only given in this section the formulas relating to the topological situations considered in §§17-18.

§17. Group actions and Rademacher reciprocity

In this section we construct two finite group actions on complex manifolds for which the equality of the G-signature theorem is precisely the generalized Rademacher reciprocity law proved in §16. To obtain the defects $\operatorname{def}(p;a_1,\dots,a_n)$, we have to look at manifolds of complex dimension $\underline{n}$; thus the original reciprocity law of Rademacher (eq. (13) of §16) corresponds to manifolds of real dimension four.

The first action is the same as that studied in §6, namely the linear action of $G = \mu_{a_0}\times\dots\times\mu_{a_n}$ on $P_n\mathbb{C}$, where μ_a denotes the group of ath roots of unity and $a_0,\dots,a_n$ are mutually prime integers. The calculation is relatively short because the fixed-point sets and their normal bundles were already found in §6.

The other action considered is the action of a finite cyclic group on a space obtained as the quotient of another finite group action on a hypersurface in complex projective space. This situation was used by Hirzebruch [17] to obtain the classical Rademacher reciprocity law from the G-signature theorem; the only difference is that for manifolds of dimension higher than four we need the whole L-class and not just the signature of certain quotient spaces, and therefore the use of the G-signature theorem must be replaced by the use of the result of §3 of Chapter I.

(I) A group action on projective space

We will use without comment the notations of §6, thus $X = P_n\mathbb{C}$, $G = \mu_{a_0}\times\dots\times\mu_{a_n}$, and $X(\zeta)$ denotes the component of X^g (for a fixed $g\in G$) given in §6(5). It follows from the assumption that the a_j's are mutually relatively prime that each $X(\zeta)$ is empty or consists of exactly one point if $\zeta \neq 1$: it is empty if ζ is different from all of the coordinates ζ_i of $\underline{g}$, and consists of the single point

$$p_i = (0:\dots:0:1:0:\dots:0) \qquad (i^{th} \text{ coordinate} = 1) \tag{1}$$

if $\zeta = \zeta_i \neq 1$. We write $\operatorname{Sign}(i)$ for the total contribution of the point p_i to $\sum_{g\in G}\operatorname{Sign}(g,X)$, i.e.

$$\operatorname{Sign}(i) = \sum_{g\in G,\ \zeta_i\neq 1} \langle L'(g,X)_{\zeta_i},[p_i]\rangle \qquad (i=0,\dots,n). \tag{2}$$

Thus $\sum_{i=0}^{n} \mathrm{Sign}(i)$ represents the total contribution to $\sum_{g\in G} \mathrm{Sign}(g,X)$ from all fixed-point set components $X(\zeta)$ with $\zeta \neq 1$. Similarly, we write S for the total contribution from the components $X(1)$, i.e.

$$S = \sum_{g\in G} \langle L'(g,X)_1, [X(1)]\rangle \tag{3}$$

Then

$$S + \sum_{i=0}^{n} \mathrm{Sign}(i) = \sum_{g\in G} \mathrm{Sign}(g,X) = |G|\ \mathrm{Sign}\ X/G, \tag{4}$$

by the usual formula for the signature of a quotient. Moreover, the action of G embeds in an action of the connected group T^{n+1} (cf. §6), so G acts trivially on $H^*(X)$ and therefore $\mathrm{Sign}\ X/G = \mathrm{Sign}\ X = \mathrm{Sign}\ P_n\mathbb{C}$. We can assume that $\underline{n}$ is even (otherwise all the signatures are zero), so that $\mathrm{Sign}\ P_n\mathbb{C} = 1$. Since $|G| = a_0\ldots a_n$, we obtain

$$S + \sum_{i=0}^{n} \mathrm{Sign}(i) = a_0\ldots a_n \qquad (\underline{n}\text{ even}). \tag{5}$$

The Rademacher reciprocity law will thus have been exhibited as a special case of the G-signature formula when we have shown that:

$$\mathrm{Sign}(i) = a_0\ldots\hat{a}_i\ldots a_n\ \mathrm{def}(a_i; a_0,\ldots,\hat{a}_i,\ldots,a_n), \tag{6}$$

$$S = L_k(p_1,\ldots,p_k) = \mathrm{res}_{y=0}\left[\prod_{j=0}^{n} \frac{a_j}{\tanh a_j y}\,dy\right]. \tag{7}$$

In the last equation we have used the notation of §16; thus $k = n/2$ and $p_j = \sigma_j(a_0^2,\ldots,a_{2k}^2)$.

To prove (6), we consider $g\in G$, $\zeta = \zeta_i \neq 1$. The eigenvalues of $\underline{g}$ on the tangent space at the isolated fixed point p_i are (cf. §6) $\zeta^{-1}\zeta_j$ $(j=0,\ldots,n,\ j\neq i)$, so

$$\langle L'(g,X)_\zeta, [p_i]\rangle = \prod_{\substack{j=0\\ j\neq i}}^{n} \frac{\zeta^{-1}\zeta_j + 1}{\zeta^{-1}\zeta_j - 1},$$

and therefore (2) becomes

$$\mathrm{Sign}(i) = \sum_{\substack{\zeta_i^{a_i}=1\\ \zeta_i\neq 1}} \prod_{\substack{j=0\\ j\neq i}}^{n} \left(\sum_{\zeta_j^{a_j}=1} \frac{\zeta_i^{-1}\zeta_j + 1}{\zeta_i^{-1}\zeta_j - 1} \right)$$

$$= \sum_{\substack{\zeta_i^{a_i}=1 \\ \zeta_i \neq 1}} \prod_{\substack{j=0 \\ j\neq i}}^{n} \left(a_j \frac{\zeta_i^{-a_i}+1}{\zeta_i^{-a_i}-1} \right), \tag{8}$$

where the last line has been obtained by a simple trigonometric identity.* The right-hand side of (8) is precisely the expression (6).

To compute S, we need to evaluate $L'(g,X)_1$ on $[X(1)]$. Renumber the coordinates (for a fixed $g = (\zeta_0,\ldots,\zeta_n) \in G$) so that $\zeta_0=\ldots=\zeta_s=1$, $\zeta_i \neq 1$ $(i=s+1,\ldots,n)$. Then $X(1) = P_s\mathbb{C}$, and if we denote the generator of $H^2(P_s\mathbb{C})$ by $\underline{y}$, we can express $L'(g,X)_1$ by

$$L'(g,X)_1 \quad = \quad \left(\frac{y}{\tanh y}\right)^{s+1} \prod_{i=s+1}^{n} \frac{\zeta_i e^{2y}+1}{\zeta_i e^{2y}-1} \tag{9}$$

(eq. (10) of §6). Evaluation on $[X(1)]$ corresponds to taking the coefficient of y^s in (9), so we obtain

$$< L'(g,X)_1,[X(1)]> \;=\; \operatorname{res}_{y=0}\left[(\coth y)^{s+1} \prod_{i=s+1}^{n} \frac{\zeta_i e^{2y}-1}{\zeta_i e^{2y}-1}\, dy \right]$$

$$= \operatorname{res}_{y=0}\left[\prod_{i=0}^{n} \frac{\zeta_i e^{2y}+1}{\zeta_i e^{2y}-1}\, dy \right]. \tag{10}$$

If we sum over all $g \in G$ for which $\zeta_0=\ldots=\zeta_s=1$, $\zeta_{s+1},\ldots,\zeta_n \neq 1$, we get as the total contribution from $P_s\mathbb{C}$ the expression

$$\operatorname{res}_{y=0}\left[(\coth y)^{s+1} \prod_{i=s+1}^{n} \left(\sum_{\substack{\zeta^{a_i}=1 \\ \zeta\neq 1}} \frac{\zeta e^{2y}+1}{\zeta e^{2y}-1} \right) dy \right]$$

$$= \operatorname{res}_{y=0} \left[(\coth y)^{s+1} \prod_{i=s+1}^{n} (\, a_i \coth a_i y \;-\; \coth y\,)\, dy \,\right], \tag{11}$$

where again the last line has been obtained by using an elementary identity. The unsymmetric form of (11) is due to the renumbering of the coordinates. In general, if I is a subset of $N = \{0,1,\ldots,n\}$, then the total contribution to S from elements $g \in G$ with $\zeta_i = 1 \Longleftrightarrow i \in I$ is

$$\operatorname{res}_{y=0} \;[\; \Pi_{i\in I}(\coth y) \;\; \Pi_{i\notin I}(\, a_i \coth a_i y \,-\, \coth y\,)\, dy \,]\,.$$

---

* Namely §6(17).

To obtain S, we must sum this over all subsets I of N, getting

$$S = \mathrm{res}_{y=0}\left[\sum_{I\subset N} \prod_{i\in I} (\coth y) \prod_{i\notin I} (a_i \coth a_i y - \coth y)\ dy \right]$$

$$= \mathrm{res}_{y=0}\left[\prod_{i\in N} (\coth y + a_i \coth a_i y - \coth y)\ dy \right],$$

which is equal to the right-hand side of equation (7).

(II) A group action on a hypersurface

We change our notation somewhat. We now let $\underline{n}$ be an odd integer and $b_1,\dots,b_n$ mutually relatively prime integers (all positive). We write N for the product $b_1\dots b_n$ and a_i for the quotient N/b_i. Let G denote the group μ_N and H the group $\mu_{b_1}\times\dots\times\mu_{b_n}$. The product G×H acts on the hypersurface

$$V = \{z = (z_0:\dots:z_n) \in P_n\mathbb{C} \mid z_0^N + \dots + z_n^N = 0\} \tag{12}$$

by

$$(\zeta,\alpha_1,\dots,\alpha_n)\circ(z_0:\dots:z_n) = (\zeta^{-1}z_0:\alpha_1 z_1:\dots:\alpha_n z_n), \tag{13}$$

where $\zeta^N=1$ and $\alpha_i^{b_i}=1$. It is easy to see that the map $V\to P_{n-1}\mathbb{C}$ given by projection onto the last $\underline{n}$ coordinates gives an isomorphism from V/G onto $P_{n-1}\mathbb{C}$. The induced action of H on $P_{n-1}\mathbb{C}$ is of the type considered above and in §6 and acts trivially on the cohomology, so

$$\mathrm{Sign}(V/G\times H) = \mathrm{Sign}(P_{n-1}\mathbb{C}/H) = \mathrm{Sign}\, P_{n-1}\mathbb{C} = 1, \tag{14}$$

the last equality holding because of the assumption that $\underline{n}$ is odd.

The quotient $W = V/H$ is naturally a complex manifold, since the action of H on V, although it does have fixed points, has at a fixed point a representation which is a sum of the standard representation of μ_b on $\mathbb{C}$, and the quotient of this standard representation is non-singular (the map $z \mapsto z^b$ gives an isomorphism $\mathbb{C}/\mu_b \to \mathbb{C}$). We will apply the G-signature theorem to the action of G on W, using

$$\mathrm{Sign}(W/G) = \mathrm{Sign}(V/G\times H) = 1. \tag{15}$$

Let Y be the hypersurface in W defined by

$$Y = \{z\in V \mid z_0 = 0\}/H. \tag{16}$$

Since G acts trivially on Y, we have $Y \subset W^g$ for every $g \in G$. We claim that, for $g \in G - \{1\}$, the fixed-point set W^g is the disjoint union of Y and finitely many isolated points. Indeed, we can identify W-Y with the Brieskorn variety

$$\{(w_1,\dots,w_n) \in \mathbb{C}^n \mid w_1^{a_1}+\dots+w_n^{a_n} = -1\} \tag{17}$$

by the map $w_i = (z_i/z_0)^{b_i}$, and in terms of these coordinates w_j for W-Y, the action of G is given by

$$\zeta \circ (w_1,\dots,w_n) = (\zeta^{b_1} w_1,\dots, \zeta^{b_n} w_n). \tag{18}$$

Therefore $\underline{w}$ is a fixed point of ζ if and only if $\zeta^{b_i} = 1$ whenever $w_i \neq 0$. This can be the case for at most one $\underline{i}$ if ζ is not the identity (since the b_i's are relatively prime), and then the condition that $\underline{w}$ lies on the variety (17) is $w_i^{a_i} = -1$. Thus ζ acts freely on W-Y if each ζ^{b_j} is different from one, while if $\zeta^{b_i} = 1$, $\zeta \neq 1$, the fixed-point set of ζ on W-Y is the set of a_i points

$$W(i) = \{(0,\dots,w_i,\dots,0) \mid w_i^{a_i} = -1\} \subset W.$$

From this description of the action of G on W, we obtain

$$\begin{aligned} N = |G| &= |G| \operatorname{Sign} W/G = \sum_{\zeta \in G} \operatorname{Sign}(\zeta,W) \\ &= \operatorname{Sign} W + S_Y + \sum_{i=1}^{n} S_i, \end{aligned} \tag{19}$$

where S_Y is the sum over $\zeta \in G - \{1\}$ of the contribution to $\operatorname{Sign}(\zeta,W)$ of the component Y of W, and S_i is similarly the total contribution from the set W(i). We claim that

$$S_i = a_i \operatorname{def}(b_i; b_1,\dots,\hat{b_i},\dots,b_n), \tag{20}$$

$$S_Y = \operatorname{res}_{y=0} [(N - \coth y \tanh Ny) \prod_{j=1}^{n} (\coth b_j y)\, dy], \tag{21}$$

$$\operatorname{Sign} W = \operatorname{res}_{y=0} [(\coth y \tanh Ny) \prod_{j=1}^{n} (\coth b_j y)\, dy]. \tag{22}$$

Clearly the Rademacher law follows from equations (19)-(22).

We begin by calculating S_i. If $w \in W(i)$ is an isolated fixed point of ζ, then we see from (18) that the eigenvalues of ζ at $\underline{w}$ are the numbers ζ^{b_j} $(1\leq j\leq n,\ j\neq i)$, and the contribution of $\underline{w}$ to $\mathrm{Sign}(\zeta,W)$ is therefore equal to

$$\prod_{\substack{j=1\\ j\neq i}}^{n} \frac{\zeta^{b_j}+1}{\zeta^{b_j}-1}\,.$$

We must multiply this with a_i (since $W(i)$ contains a_i points, all with the same eigenvalues) and sum over all $\zeta\in G$ for which $W^{\zeta} = Y\cup W(i)$, i.e. over all ζ with $\zeta^{b_i}=1$, $\zeta\neq 1$. This proves equation (20).

To prove (21) and (22), we will need to calculate the L-class of W. First we evaluate the L-class of the hypersurface (12). Let $\underline{i}$ denote the inclusion of V in $P_n\mathbb{C}$ and ν the normal bundle, and let $x\in H^2(P_n\mathbb{C})$ be the standard generator. Clearly ν is a complex line bundle with $c_1(\nu) = Ny$, where $y = i^*x \in H^2(Y)$. Therefore

$$L(V) \;=\; L(\nu)^{-1}\, i^*L(P_n\mathbb{C}) \;=\; \frac{\tanh Ny}{Ny}\left(\frac{y}{\tanh y}\right)^{n+1}. \tag{23}$$

To calculate the L-class of $W = Y/H$, we will use the theorem of §3. We must first find the fixed-point sets. Let $\alpha = (\alpha_0,\dots,\alpha_n)\in H$ (here $\alpha_0=1$, i.e. we have identified H with $1\times H$ for convenience; thus $\alpha_i^{b_i} = 1$ where $b_0 = 1$). Using the results of §6, we find

$$V^{\alpha} \;=\; (P_n\mathbb{C})^{\alpha}\cap V \;=\; \bigcup_{\zeta\in S^1}\{\, z\in V \mid \alpha_i\neq\zeta \Rightarrow z_i = 0\,\}$$

$$=\; \{z\in V \mid \alpha_i\neq 1 \Rightarrow z_i=0\}. \tag{24}$$

The last equality follows because the integers $1,b_1,\dots,b_n$ are coprime, so for $\zeta\in S^1-\{1\}$ there can be at most one $\underline{i}$ with $\alpha_i=\zeta$, whereas V does not contain any point with only one non-zero coordinate. If we renumber the coordinates so that $\alpha_0=\dots=\alpha_s=1$, $\alpha_{s+1},\dots,\alpha_n\neq 1$, we find

$$V^{\alpha} \;=\; \{(z_0:\dots:z_s:0:\dots:0)\in P_n\mathbb{C} \mid z_0^N+\dots+z_s^N = 0\}, \tag{25}$$

so that the fixed-point set of $\alpha\in H$ on V is also a hypersurface of degree N. Clearly its normal bundle in V is the direct sum of n-s copies of the restriction $i^*\eta$ of the Hopf bundle η over $P_s\mathbb{C}$ to V^{α}, and the action of α on this normal bundle consists of multiplication with α_{s+j} $(j=1,\dots,n-s)$ on the jth summand. We denote by $\hat{x}$ the standard generator of $H^2(P_s\mathbb{C})$ and by $\hat{y}= i^*\hat{x}$ its restriction to V^{α}. Then, in view of (23) and of the description just given of the normal bundle of V^α in V, we obtain from the G-signature theorem that

$$L'(\alpha,V) = \frac{\tanh N\hat{y}}{N\hat{y}} \left(\frac{\hat{y}}{\tanh \hat{y}}\right)^{s+1} \prod_{j=s+1}^{n} \frac{\alpha_j e^{2\hat{y}}+1}{\alpha_j e^{2\hat{y}}-1} . \tag{26}$$

It is also easy to see that $j_!(\hat{y}^r) = y^{n-s+r}$, where $\underline{j}$ is the inclusion map from V^α to V; therefore

$$L(\alpha,V) = j_! L'(\alpha,V) = \frac{\tanh Ny}{Ny} \prod_{j=0}^{n} \left(y \frac{\alpha_j e^{2y}+1}{\alpha_j e^{2y}-1} \right) . \tag{27}$$

Summing this over $\alpha \epsilon H$ (recall that $\alpha_0 = 1$) and using the theorem of §3, we obtain

$$\pi^* L(W) = \frac{\tanh Ny}{Ny} \prod_{j=0}^{n} \left(y \sum_{\alpha^{b_j}=1} \frac{\alpha e^{2y}+1}{\alpha e^{2y}-1} \right)$$

$$= \frac{\tanh Ny}{Ny} \prod_{j=0}^{n} \frac{b_j y}{\tanh b_j y} , \tag{28}$$

where π denotes the projection from V to V/H = W, $b_0 = 1$, and the last line has been obtained by the usual identity.

It follows immediately that (22) holds. Indeed,

$$\text{Sign } W = \langle L(W),[W]\rangle = \frac{1}{|H|} \langle L(W),\pi_*[V]\rangle = \frac{1}{|H|} \langle \pi^* L(W),[V]\rangle$$

$$= \frac{1}{|H|} \left(\text{coefficient of } \frac{y^{n-1}}{N} \text{ in } \pi^* L(W) \right)$$

(this last equation holds because $i_*[V] \epsilon H_{2n-2}(P_n\mathbb{C})$ is the Poincaré dual of $Nx \epsilon H^2(P_n\mathbb{C})$, and evaluation on the fundamental class in $P_n\mathbb{C}$ consists in picking out the coefficient of x^n), and since $N = |H| = \Pi_{j=0}^{n} b_j$, we obtain equation (22). But the calculation (28) can also be used to evaluate the number S_Y appearing in (21). Indeed, Y is defined exactly like W but with one coordinate fewer, so we have

$$L(Y) = \frac{\tanh Nz}{Nz} \prod_{j=1}^{n} \frac{b_j z}{\tanh b_j z} , \tag{29}$$

where $z \epsilon H^2(Y)$ is the analogue a dimension lower of $(\pi^*)^{-1} y \epsilon H^2(W)$ (we use complex coefficients where π^* is an isomorphism). The complex manifold Y is embedded in W with a normal bundle whose first Chern class is precisely $\underline{z}$, and $G = \mu_N$ acts on this normal line bundle by multiplication with N^{th} roots of unity. Finally, evaluation on [Y] corresponds to finding the coefficient of y^{n-2}/N. Therefore the G-signature theorem tells us that

$$S_Y = \sum_{\substack{\zeta^N=1 \\ \zeta\neq 1}} \left< L(Y) \frac{\zeta e^{2z}+1}{\zeta e^{2z}-1}, [Y] \right>$$

$$= \operatorname{res}_{z=0} \left[\frac{N}{z^{n-2}} \frac{dz}{z} \frac{\tanh Nz}{Nz} \prod_{j=1}^{n} \left(\frac{b_j z}{\tanh b_j z} \right) \sum_{\substack{\zeta^N=1 \\ \zeta\neq 1}} \frac{\zeta e^{2z}+1}{\zeta e^{2z}-1} \right]$$

$$= \operatorname{res}_{z=0} \left[dz \left(\tanh Nz \prod_{j=1}^{n} \coth b_j z \right) (N \coth Nz - \coth z) \right] .$$

This completes the proof of equation (21).

§18. Equivariant signature of Brieskorn varieties

Let $a = (a_1,\dots,a_n)$ be an n-tuple of integers ≥ 2, and

$$V_a = \{z \in \mathbb{C}^n \mid z_1^{a_1}+\dots+z_n^{a_n} = 1\} \tag{1}$$

be the corresponding Brieskorn variety. The group

$$G = \mu_{a_1}\times\dots\times\mu_{a_n} \tag{2}$$

(where μ_a is the group of ath roots of unity) acts on V_a by

$$\zeta\circ z = (\zeta_1 z_1,\dots,\zeta_n z_n) \qquad (\zeta \in G,\ z \in V_a). \tag{3}$$

Here ζ_i or z_i denotes the ith component of ζ, z respectively.

We will evaluate the equivariant signatures for this action by using the known description of the cohomology of V_a and the action of G on this cohomology. We state the general result as a theorem, and single out three special cases as corollaries. The first, obtained by taking $\zeta = \mathrm{id}$, is Brieskorn's result for the signature of V_a itself. The second is a result of Hirzebruch and Jänich on the Browder-Livesay invariant of a certain involution on V_a defined when each a_j is even. The third gives the equivariant signature for the action of a certain cyclic group embedded in G; when this action is free it embeds in a free S^1-action the signature of which will be calculated later in this section.

To state the theorem, we use abbreviated notation for n-tuples. Thus if $\zeta = (\zeta_1,\dots,\zeta_n) \in G$ and $j = (j_1,\dots,j_n)$ is an n-tuple of integers, we write ζ^j for $\zeta_1^{j_1}\dots\zeta_n^{j_n}$. Similarly $0<j<a$ means that $0<j_k<a_k$ $(k=1,\dots,n)$ and $\frac{j}{a}$ denotes $\frac{j_1}{a_1}+\dots+\frac{j_n}{a_n}$.

Theorem 1: The signature of $\zeta = (\zeta_1,\dots,\zeta_n) \in G$ on V_a is given by

$$\mathrm{Sign}(\zeta,V_a) = \sum_{0<j<a} \varepsilon\left(\tfrac{j}{a}\right)\zeta^j, \tag{4}$$

where

$$\varepsilon(x) = \begin{cases} +1, & \text{if } 0<x<1 \pmod 2,\\ -1, & \text{if } 1<x<2 \pmod 2,\\ 0, & \text{if } x\in\mathbb{Z}. \end{cases} \tag{5}$$

This can also be written as a trigonometric sum:

$$\operatorname{Sign}(\zeta, V_a) = \frac{1}{N} \sum_{t^N=-1} \frac{1+t}{1-t} \prod_{k=0}^{n} \frac{1+\zeta_k t^{-b_k}}{1-\zeta_k t^{-b_k}} \qquad (6)$$

$$= \frac{i^{n+2}}{N} \sum_{\substack{j=1 \\ j \text{ odd}}}^{2N-1} \cot\frac{\pi j}{2N} \cot\frac{\pi(2s_1-j)}{2a_1} \cdots \cot\frac{\pi(2s_n-j)}{2a_n}. \qquad (7)$$

Here N is any common multiple of $a_1,\dots,a_n$ and $b_k = N/a_k$. The numbers s_k in equation (7) are defined (mod a_k) by

$$\zeta_k = e^{2\pi i s_k/a_k} . \qquad (8)$$

Corollary 1 (Brieskorn [3]): The signature of V_a is zero for n odd and, for n even is given by

$$\operatorname{Sign}(V_a) = \sum_{0<j<a} \varepsilon(\tfrac{j}{a}) = t(a) \qquad (9)$$

$$= \frac{(-1)^{\frac{n}{2}+1}}{N} \sum_{\substack{j=1 \\ j \text{ odd}}}^{2N-1} \cot\frac{j\pi}{2N} \cot\frac{j\pi}{2a_1} \cdots \cot\frac{j\pi}{2a_n} . \qquad (10)$$

Corollary 2 (Hirzebruch and Jänich [19]): If $a_1,\dots,a_n$ are even, the involution $T: V_a \to V_a$ sending $(z_1,\dots,z_n)$ to $(-z_1,\dots,-z_n)$ has signature

$$\operatorname{Sign}(T, V_a) = \sum_{0<j<a} \varepsilon(\tfrac{j}{a}) \cdot (-1)^{j_1+\dots+j_n} \qquad (11)$$

$$= \frac{(-1)^{\frac{n}{2}}}{N} \sum_{\substack{j=1 \\ j \text{ odd}}}^{2N-1} \cot\frac{j\pi}{2N} \tan\frac{j\pi}{2a_n} \cdots \tan\frac{j\pi}{2a_n} . \qquad (12)$$

Corollary 3: Let N, b_k be as above. The μ_N acts on V_a by

$$t \circ (z_1,\dots,z_n) = (t^{b_1} z_1,\dots,t^{b_n} z_n) \qquad (t^N=1). \qquad (13)$$

The signature of this action is given by

$$\operatorname{Sign}(e^{2\pi i h/N}, V_a) = \sum_{0<j<a} \varepsilon(\tfrac{j}{a})\, e^{2\pi i h(j/a)} \qquad (14)$$

$$= \frac{i^{n+2}}{N} \sum_{\substack{j=1 \\ j \text{ odd}}}^{2N-1} \cot\frac{\pi(2h-j)}{2N} \cot\frac{\pi j}{2a_1} \cdots \cot\frac{\pi j}{2a_n} . \qquad (15)$$

Proof: The corollaries are obtained from the theorem by specializing ζ

and performing simple manipulations with the formulas. The proof that (4) and (6) are equal is exactly like the proof given in §16 for the special case $\zeta = \mathrm{id}$ (i.e. for the equality of (9) and (10)). We thus only need to prove equation (4). To do so, we use the results on the cohomology of V_a given in Pham [35], Brieskorn [3], Hirzebruch-Mayer [20].

One can give a G-equivariant deformation retraction of V_a onto

$$U_a = \{z \in V_a \mid z_k^{a_k} \text{ is real and } \geq 0 \ (k=1,\ldots,n)\}. \tag{16}$$

The space U_a is naturally homeomorphic to the join $\mu_{a_1} * \ldots * \mu_{a_n}$ of the discrete spaces μ_{a_j} $(j=1,\ldots,n)$. It follows that V_a is $(n-2)$-connected and $H_{n-1}(V_a)$ is free abelian of rank $\prod_k (a_k-1)$. We denote by w_k the action of the generator of μ_{a_k} on $H_*(V_a)$. Let $\underline{e}$ be the $(n-1)$-simplex of U_a that corresponds to $1 \in G$ in the identification of U_a with $\mu_{a_1} * \ldots * \mu_{a_n}$. The group $C_{n-1}(U_a)$ of n-chains is $\mathbb{Z}[G]e$, where $\mathbb{Z}[G]$ is the group ring of G. Since the boundary operator from C_{n-1} to C_{n-2} commutes with the action of G, the element

$$h = (1-w_1)\ldots(1-w_n)e \in \mathbb{Z}[G] \tag{17}$$

is a cycle. In fact $\underline{h}$ is a generator of $H_{n-1}(U_a)$ as a $\mathbb{Z}[G]$-module:

$$H_{n-1}(U_a) \cong \mathbb{Z}[G]h \cong \mathbb{Z}[G]/(1+w_k+\ldots+w_k^{a_k-1},\ 1 \leq k \leq n). \tag{18}$$

This describes $H_{n-1}(U_a)$. We can take as a basis the set of monomials $w^j = w_1^{j_1}\ldots w_n^{j_n}$ with $0<j<a$, and the action of G is given in the obvious way by taking into account the relations $1+w_k+\ldots+w_k^{a_k-1} = 0$. Finally, the intersection form on $H_{n-1}(U_a)$ is given by

$$S(xh,yh) = E(x\bar{y}\eta) \qquad (x,y \in \mathbb{Z}[G]), \tag{19}$$

where $\eta = (1-w_1)\ldots(1-w_n) \in \mathbb{Z}[G]$ and E is defined on the generators of $\mathbb{Z}[G]$ by

$$E(w_1^{j_1}\ldots w_n^{j_n}) = \begin{cases} \delta, & \text{if } j_1=\ldots=j_n=0, \\ -\delta, & \text{if } j_1=\ldots=j_n=1, \\ 0, & \text{otherwise,} \end{cases} \tag{20}$$

where $\delta = (-1)^{n(n-1)/2}$. For $r = (r_1,\ldots,r_n)$ an r-tuple of integers defined modulo $\underline{a}$, we denote by L_r the element $w^r\mathbf{h} = w_1^{r_1}\ldots w_n^{r_n}\mathbf{h}$ of $H_{n-1}(U_a)$, or rather the corresponding element in $H_{n-1}(V_a)$ under the isomorphism induced by the inclusion of U_a in V_a. Then (19) implies:

$$S(L_r,L_s) = \begin{cases} (-1)^t\delta & \text{if } \underline{t} \text{ of the numbers } s_k-r_k \text{ are } =1 \text{ and the rest } =0, \\ -(-1)^t & \text{if } \underline{t} \text{ of the numbers } s_k-r_k \text{ are } =0 \text{ and the rest } =-1, \\ 0 & \text{otherwise.} \end{cases} \tag{21}$$

We define a new basis for $H_{n-1}(V_a)$ by

$$v_j = \prod_{k=1}^{n} (1 + \xi_k^{j_k} w_k + \dots + \xi_k^{(a_k-1)j_k} w_k^{a_k-1})\, h \tag{22}$$

$$= \sum_{r \bmod a} \xi^{jr} L_r , \tag{23}$$

where $\xi_k = e^{2\pi i/a_k}$, $j = (j_1,\dots,j_n)$ is an n-tuple of integers $j_k \not\equiv 0$ (mod a_k), and ξ^{jr} denotes $\xi_1^{j_1r_1}\dots\xi_n^{j_nr_n}$. Then, from (21) and (23),

$$S(v_j,v_k) = \sum_{r,s \bmod a} \xi^{jr+ks}\, S(L_r,L_s)$$

$$= \delta \sum_{r \bmod a} \xi^{(j+k)r} [(1-\xi_1^{k_1})\dots(1-\xi_n^{k_n}) - (-1)^n(1-\xi_1^{-k_1})\dots(1-\xi_n^{-k_n})]$$

$$= \delta\, (1-\xi_1^{k_1})\dots(1-\xi_n^{k_n})(1-\xi_1^{-k_1}\dots\xi_n^{-k_n}) \prod_{i=1}^{n} \Big(\sum_{r=1}^{a_i} \xi_i^{(j_i+k_i)r_i} \Big)$$

$$= \begin{cases} \delta a_1\dots a_n(1-\xi_1^{k_1})\dots(1-\xi_n^{k_n})(1-\xi_1^{-k_1}\dots\xi_n^{-k_n}) & \text{if } j+k\equiv 0 \pmod a, \\ 0 & \text{otherwise.} \end{cases}$$

Thus the only non-zero elements of the matrix of S with respect to the v_j (where we fix $\underline{j}$ by $0<j<a$) are the elements

$$c_j = S(v_j,v_{-j}) = \delta\, a_1\dots a_n \prod_{i=1}^{n}(\xi_i^{j_i/2} - \xi_i^{-j_i/2})(\xi_1^{-j_1/2}\dots\xi_n^{-j_n/2} - \xi_1^{j_1/2}\dots\xi_n^{-j_n/2})$$

$$= -(-1)^{n(n-1)/2}\, (2i)^{n+1} \sin\Big(\frac{\pi j_1}{a_1}+\dots+\frac{\pi j_n}{a_n}\Big) \sin\frac{\pi j_1}{a_1}\dots\sin\frac{\pi j_n}{a_n} \tag{24}$$

$$= i^{n^2+1}\, \varepsilon\big(\tfrac{j}{a}\big) \text{ (real, positive number).} \tag{25}$$

In particular, S is a non-degenerate form if and only if the n-tuple $\underline{a}$ is such that $\frac{j}{a} = \frac{j_1}{a_1} + \dots + \frac{j_n}{a_n} \notin \mathbb{Z}$ whenever $0<j<a$.

We introduce a new basis

$$A_j = v_j + v_{a-j}, \quad B_j = i(v_j - v_{a-j}), \qquad (j \in M), \qquad (26)$$

where M is a set of indices $\underline{j}$ with $0<j<a$ such that, for any $\underline{k}$ with $0<k<a$, exactly one of k, $a-k$ is in M (this is necessary to have a basis, since the elements (26) satisfy $A_{a-j} = A_j$, $B_{a-j} = -B_j$).

We now suppose that $\underline{n}$ is **odd**. Then the intersection form M is symmetric, and we obtain $(j,k \in M)$

$$S(A_j,A_k) = 2\,\delta_{jk}\,c_j, \quad S(B_j,B_k) = 2\,\delta_{jk}\,c_j, \quad S(A_j,B_k) = 0, \qquad (27)$$

where δ_{jk} is the Kronecker delta. Thus S is diagonal. If we write

$$\zeta = (\zeta_1,\dots,\zeta_n) = w_1^{s_1}\dots w_n^{s_n} \in G \qquad (28)$$

(thus $\zeta_k = e^{2\pi i s_k/a_k}$; the notation using the w's is the one introduced to denote the action of G on $H_{n-1}(V_a)$), then it is clear from (23) that v_j is an eigenvalue of the action of ζ in homology:

$$\zeta_* v_j = \zeta^{-j} v_j = e^{i\theta(j)} v_j, \qquad (29)$$

where ζ^{-j} has the meaning explained before Theorem 1 and

$$\theta(j) = 2\pi \left(\frac{s_1 j_1}{a_1} + \dots + \frac{s_n j_n}{a_n}\right). \qquad (30)$$

Therefore

$$\zeta_* A_j = A_j \cos\theta(j) - B_j \sin\theta(j), \quad \zeta_* B_j = B_j \cos\theta(j) + A_j \sin\theta(j).$$

If we substitute this and equations (27), (25) into the definition of the equivariant signature for a manifold of dimension $\equiv 0 \pmod 4$, we get

$$\mathrm{Sign}(\zeta,V_a) = \mathrm{tr}(\zeta_*|H_{n-1}(V_a)) - \mathrm{tr}(\zeta_*|H_{n-1}(V_a))$$

$$= \sum_{j\in M} 2\,\varepsilon(j)\cos\theta(j).$$

Since $\varepsilon(j)$ and $\cos\theta(j)$ are the same for $\underline{j}$ and $a-j$, we can write this

$$\mathrm{Sign}(\zeta,V_a) = \sum_{0<j<a} \varepsilon(j)\cos\theta(j).$$

This is equivalent to (4) since $\underline{n}$ is odd.

Now assume that $\underline{n}$ is even. Then, for $j,k \in M$, we have

$$S(A_j,A_k) = S(B_j,B_k) = 0, \quad S(A_j,B_k) = -S(B_k,A_j) = -2i\,\delta_{jk}\,c_j. \tag{28}$$

Therefore the matrix $C:H_{n-1}(V_a) \to H_{n-1}(V_a)$ corresponding to S after the introduction of the scalar product defined by the basis $\{A_j, B_j \mid j \in M\}$ is $\begin{pmatrix} 0 & -2ic \\ 2ic & 0 \end{pmatrix}$ with respect to this basis, where $\underline{c}$ is the diagonal matrix of the c_j's $(j \in M)$. Therefore $CC^* = -C^2$ has the matrix $\begin{pmatrix} -4c^2 & 0 \\ 0 & -4c^2 \end{pmatrix}$ with respect to his basis; this is positive definite since c_j is pure imaginary for even $\underline{n}$ by (25). The positive definite square root of this has matrix $\begin{pmatrix} 2|c| & 0 \\ 0 & 2|c| \end{pmatrix}$. Therefore the square root of $-I$ appearing in the definition of $\mathrm{Sign}(\zeta,V)$ for a manifold V of dimension $\equiv 2 \pmod 4$ (see (III) of §2) has the matrix

$$J = C/(CC^*)^{1/2} = \begin{pmatrix} 0 & -ic/|c| \\ ic/|c| & 0 \end{pmatrix}, \tag{29}$$

that is, J is given by

$$J(A_j) = -\frac{ic_j}{|c_j|}B_j, \quad J(B_j) = \frac{ic_j}{|c_j|}A_j. \tag{30}$$

If we substitute (25) for c_j and observe that $i^{n^2-1} = -i$ for even $\underline{n}$, we get

$$J(A_j) = -\varepsilon(j)\,B_j, \quad J(B_j) = \varepsilon(j)\,A_j. \tag{31}$$

In general, if V is a real vector space on which $J:V \to V$ is a map with square $-I$, and V has a basis of the form $e_1,\dots,e_r,Je_1,\dots,Je_r$, then for a map $G:V \to V$ commuting with J, the trace of G thought of as a complex matrix on a complex vector space is $\operatorname{tr} G' + i \operatorname{tr} G''$, where G' and G'' are the matrices defined by $G(e_j) = \sum_k (G'_{jk}e_k + G''_{jk}Je_k)$. We take for V the homology group $H_{n-1}(V_a)$ and for G the action of ζ. For $e_1,\dots,e_r$ we take $\{A_j \mid j \in M\}$. The action of ζ is just as in the case of odd $\underline{n}$, i.e.

$$\begin{aligned} \zeta_* A_j &= A_j \cos\theta(j) - B_j \sin\theta(j) \\ &= \cos\theta(j)\,A_j + \varepsilon(j)\sin\theta(j)\,JA_j, \end{aligned}$$

and we conclude that

$$\mathrm{Sign}(\zeta,V_a) = 2i\,\mathrm{Im}(\operatorname{tr}(\zeta \mid H_{n-1}(V_a)_J)$$

$$= 2i \sum_{j \in M} \varepsilon(j) \sin \theta(j).$$

Since $\varepsilon(a-j) = -\varepsilon(j)$, $\sin \theta(a-j) = -\sin \theta(j)$ for even $\underline{n}$, this equals

$$i \sum_{0<j<a} \varepsilon(j) \sin \theta(j) = \sum_{0<j<a} \varepsilon(j)\, e^{i\theta(j)}.$$

This proves (4) for even $\underline{n}$ and completes the proof of Theorem 1.

We now wish to study the odd-dimensional smooth manifold

$$\Sigma_a = \{z \in \mathbb{C}^n \mid z_1^{a_1}+\ldots+z_n^{a_n} = 0,\ |z_1|^2+\ldots+|z_n|^2 = 1\}. \tag{32}$$

This is diffeomorphic to the boundary of $V_a \cap D$, where D is a disc in $\mathbb{C}^n$ of large radius. The advantage of using the homogeneous equation $\Sigma z_k^{a_k} = 0$ is that the action of μ_N described in Corollary 3 of Theorem 1 now extends to an S^1-action, defined by the same formula (13) with $t \in S^1$.

Recall that the G-signature theorem can be used to define an invariant of certain group actions on odd-dimensional manifolds (Atiyah and Singer [1]). The definition of this "α-invariant" is as follows: if the disjoint union of $\underline{m}$ copies of a G-manifold Σ is equivariantly diffeomorphic to the boundary of a G-manifold X, then

$$\alpha(t,\Sigma) = \frac{1}{m} \{ \operatorname{Sign}(t,X) - L'(t,X)[X^t] \} \tag{33}$$

for any $t \in G$ acting freely on Σ (so that $X^t \cap \partial X = \emptyset$). It has also been proved (Ossa [34]) that, if $G = S^1$ and the action on Σ is fixed-point free (i.e. G_x is a finite subgroup of S^1 for every $x \in \Sigma$), then some multiple $\underline{m}$ of Σ always does bound an S^1-manifold (we can even take $\underline{m}$ to be a power of two), and therefore the α-invariant is defined. It is a rational function of $\underline{t}$ which can only have poles at values of $\underline{t}$ which have fixed-points on Σ. The action of S^1 on Σ_a given by (13) is certainly fixed-point free; indeed, it is clear that $t \in S^1$ can only have a fixed point if $t^N = 1$. However, we will not be able to calculate the α-invariant of this action on the Brieskorn manifold in general. The problem is to find the manifold X with $m\Sigma_a = \partial X$; we can take m=1 and $X = V_a \cap D$ as above, but the diffeomorphism of Σ_a onto $V_a \cap \partial D$ can only be made μ_N-equivariant since there is no natural action of S^1 on V_a.

There is, however, one case for which we can calculate the α-invariant, namely when the action (13) is free. This is a general fact: if Σ is a free S^1-manifold, then the projection $\Sigma \to \Sigma/S^1 = Z$

defines an S^1-bundle ξ over Z, and we can consider Σ as the boundary of the associated D^2-bundle X. Clearly $X^t = Z$ for $t \in S^1 - \{1\}$. Then equation (33) gives (Atiyah and Singer [1])

$$\alpha(t,\Sigma) = \text{Sign}\ \varphi - \left\langle \frac{te^{2x}+1}{te^{2x}-1} L(Z),[Z] \right\rangle , \tag{34}$$

where $x \in H^2(Z)$ is the first Chern class of the complex line bundle ξ and φ is the following quadratic form on $H^{n-3}(Z)$ (where $2n-3 = \dim \Sigma$):

$$\varphi(\alpha,\beta) = \langle \alpha\beta x, [Z] \rangle \qquad (\alpha,\beta \in H^{n-3}(Z)) . \tag{35}$$

We now apply this with $\Sigma = \Sigma_a$ (we correspondingly write X_a and Z_a for the X and Z above). We first have to know when the action of the group S^1 on $_a$ is free; this is the case if the n numbers $b_k = N/a_k$ are mutually coprime. The quotient $Z_a = \Sigma_a/S^1$ is then a complex manifold. We will need the following facts about Z_a and the class

$$x = c_1(\xi) \in H^2(Z_a), \tag{36}$$

which were communicated to the author by W. D. Neumann:

Theorem 2 (Neumann, unpublished): The cohomology of Z_a is

$$H^i(Z_a) = \begin{cases} \mathbb{Z} & \text{if } i \text{ is even} \\ 0 & \text{if } i \text{ is odd} \end{cases} \oplus \begin{cases} H^{n-1}(\Sigma_a) & \text{if } i=n-2 \\ 0 & \text{if } i \neq n-2 \end{cases} \tag{37}$$

(here $H^{n-1}(\Sigma_a)$ is free abelian of rank $|\{j \mid 0 < j < a, \frac{j}{a} \in \mathbb{Z}\}|$; this is a consequence of the description of the intersection form (19)) and the generator of the summand $\mathbb{Z}$ in dimension 2k, denoted γ_k, is related to the element (36) by

$$x^k = \begin{cases} \gamma_k & \text{if } 2k < n-2, \\ d\,\gamma_k & \text{if } 2k \geq n-2. \end{cases} \tag{38}$$

Here $d = N/b_1 \dots b_n$, which is an integer since each b_k divides N and the b_k's are mutually coprime. The Chern class of $T(Z_a)$ equals

$$\frac{1}{1+Nx} \prod_{k=1}^{n} (1 + b_k x). \tag{39}$$

Using this information, we can easily evaluate the right-hand side of (34). It follows immediately from (39) that the L-class of Z_a is

$$L(Z_a) = \frac{\tanh Nx}{Nx} \prod_{k=1}^{n} \frac{b_k x}{\tanh b_k x} . \tag{40}$$

Furthermore, we have

$$x^{n-2}[Z_a] = -d, \tag{41}$$

and therefore the second term in (34) equals

$$\begin{aligned} \alpha(t,\Sigma_a) - \mathrm{Sign}\,\varphi &= d \cdot \text{coefficient of } x^{n-2} \text{ in } \frac{te^{2x}+1}{te^{2x}-1} L(Z_a) \\ &= d \cdot \mathrm{res}_{x=0}\left[\frac{dx}{x^{n-1}}\ \frac{te^{2x}+1}{te^{2x}-1}\ \frac{\tanh Nx}{Nx} \prod_{k=1}^{n} \frac{b_k x}{\tanh b_k x}\right] \\ &= \mathrm{res}_{x=0}\left[\frac{te^{2x}+1}{te^{2x}-1} \tanh Nx \prod_{k=1}^{n} \coth b_k x \ \ dx\right] \\ &= \mathrm{res}_{z=1}\left[\frac{tz+1}{tz-1}\ \frac{z^N-1}{z^N+1} \prod_{k=1}^{n} \frac{z^{b_k}+1}{z^{b_k}-1}\ \frac{dz}{2z}\right], \end{aligned} \tag{42}$$

where the last line has been obtained by substituting $z = e^{2x}$. The rational function in square brackets in (42) has, as well as the pole at $z=1$, poles at 0, ∞, t^{-1}, and values of $\underline{z}$ with $z^N=-1$. The factors $z^{b_k}-1$ in the denominator do not give new poles, since the zeroes of these polynomials for different $\underline{k}$ are (except for $z=1$) all distinct because the b_k's are relatively prime, and the simple zero in the denominator at a point $z^{b_k} = 1$, $z\neq1$ is offset by the vanishing of z^N-1 in the numerator at such a point. All the poles except $z=1$ are simple and their residues therefore easy to evaluate. Applying the residue theorem then gives

$$\begin{aligned} \alpha(t,\Sigma_a) - \mathrm{Sign}\,\varphi &= -\mathrm{res}_0 - \mathrm{res}_\infty - \mathrm{res}_{t^{-1}} - \sum_{z^N=-1} \mathrm{res}_z \\ &= \frac{-(-1)^n}{2} + \frac{1}{2} - \frac{t^{-N}-1}{t^{-N}+1} \prod_{k=1}^{n} \frac{t^{-b_k}+1}{t^{-b_k}-1} - \frac{1}{N} \sum_{z^N=-1} \frac{tz+1}{tz-1} \prod_{k=1}^{n} \frac{z^{b_k}+1}{z^{b_k}-1} . \end{aligned} \tag{43}$$

To evaluate Sign φ, we use equation (37). We find that $H^{n-3}(Z_a)$ is zero if $\underline{n}$ is even, while if $\underline{n}$ is odd it is isomorphic to $\mathbb{Z}$ with generator $x^{(n-3)/2}$. Since $x^{n-2}[Z_a]$ is negative (eq. (41)), the signature of the quadratic form (35) is -1 in the latter case; thus

$$\mathrm{Sign}\,\varphi = \frac{(-1)^n - 1}{2} . \tag{44}$$

Combining (43), and (44) and using the usual identity

$$\frac{t^N - 1}{t^N + 1} = \frac{1}{N} \sum_{z^N = -1} \frac{zt + 1}{zt - 1}, \tag{45}$$

we obtain:

Theorem 3: Let $b_1, \ldots, b_n$ be positive, mutually coprime integers, $\underline{d}$ a positive integer, $N = db_1 \ldots b_n$, and $a_k = N/b_k$. The free S^1-action

$$t \circ (z_1, \ldots, z_n) = (t^{b_1} z_1, \ldots, t^{b_n} z_n) \qquad (t \in S^1)$$

on the Brieskorn manifold (32) has α-invariant (for $t \neq 1$) given by

$$\alpha(t, \Sigma_a) = \frac{1}{N} \sum_{z^N = -1} \frac{t+z}{t-z} \left[\prod_{k=1}^{n} \frac{1 + t^{b_k}}{1 - t^{b_k}} - \prod_{k=1}^{n} \frac{1 + z^{b_k}}{1 - z^{b_k}} \right]. \tag{46}$$

Notice that the values $t = z$ in (46) do not give poles since then the two products in the square brackets agree. The values $t^{b_k} = 1$, $t \neq 1$, also do not give poles of $\alpha(t)$, as one can see from the alternate expression (43) (by the argument given above since the b_k's are coprime and divide N). Therefore (46) defines a rational function of $\underline{t}$ whose only pole is at $t=1$, which is as it should be for the α-invariant of a free circle action.

To connect this result with the signature calculations at the beginning of this section, we observe that, for $t \in \mu_N - \{1\}$, we can calculate the α-invariant by using the diffeomorphism from Σ_a to $\partial(V_a \cap D)$, since this diffeomorphism is μ_N-equivariant. Then $m=1$ in (33) and $\mathrm{Sign}(t,X) = \mathrm{Sign}(t,V_a)$ is the number calculated in Corollary 3 of Theorem 1, so we only have to calculate $L'(t,V_a)[V_a^t]$ (we can replace $V_a \cap D$ by V_a everywhere if the radius of D is large enough). Because the b_k's are coprime, this is easy: the fixed-point set of $t \neq 1$ is empty unless $t^{b_k} = 1$ for some $\underline{k}$ and consists of a_k isolated points with eigenvalues t^{b_j} $(j=1, \ldots, \hat{k}, \ldots, n)$ if $t^{b_k}=1$. The calculation is the one done in (II) of §17. We find that, for $t^N=1$, $t \neq 1$, the value of $-L'(t,V_a)[V_a^t]$ is precisely the value of the first sum in (46) (i.e. of the sum of the first product in the square brackets), while from (15) we see that $\mathrm{Sign}(t,V_a)$ is just the second sum in (46). This provides a check on the calculations and, incidentally, an alternate proof of the equality of (14) and (15), at least when the numbers $\frac{N}{a_k}$ are coprime.

REFERENCES

1. M. F. Atiyah and I. M. Singer: The index of elliptic operators: III, Ann. of Math. 87 (1968) 546-604

2. A. Borel: Seminar on transformation groups, Ann. of Math. Studies 46, Princeton University Press 1960

3. E. Brieskorn: Beispiele zur Differentialtopologie von Singularitäten, Inventiones Math. 2 (1966) 1-14

4. K. Chandrasekharan: Arithmetical functions, Grundlehren der math. Wiss. 167, Springer-Verlag, Berlin-Heidelberg-New York 1970

5. R. Dedekind: Erläuterungen zu zwei Fragmenten von Riemann, Riemann's gesammelte mathematische Werke und wissenschaftlicher Nachlaß, 2. Auflage (ed. H. Weber) 1892 (Dover publications, New York 1953, 466-478)

6. U. Dieter: Beziehungen zwischen Dedekindschen Summen, Abhand. Math. Sem. Hamburg 21 (1957) 109-125

7. U. Dieter: Pseudo-random numbers: the exact distribution of pairs, Math. of Computation 25 (1971)

8. A. Dold: (Co-)homology properties of topological manifolds, Conference on the topology of manifolds (ed. J. Hocking), Prindle, Weber and Schmidt, Boston, 1968, 47-58

9. A. Dold and R. Thom: Une généralisation de la notion d'espace fibré. Applications aux produits symmétriques infinis, C. R. Acad. Sci. Paris 242 (1956) 1680-1682

10. A. Grothendieck: Sur quelques points d'algèbre homologique, Tohoku Math. J. 9 (1957) 119-221

11. J. Heithecker: Homologietheorie mit lokalen Koeffizienten und einige Anwendungen, Bonn Diplomarbeit, 1970

12. F. Hirzebruch: Topological methods in algebraic geometry, third enlarged edition, Springer-Verlag, Berlin-Heidelberg-New York, 1966

13. F. Hirzebruch: Elliptische Differentialoperatoren auf Mannigfaltigkeiten, Arbeitsgemeinschaft für Forschung des Landes Nordrhein-Westfalen 33 (1965) 563-608

14. F. Hirzebruch: Lectures on the Atiyah-Singer theorem and its applications, Berkeley, Summer 1968 (unpublished lecture notes)

15. F. Hirzebruch: The signature of ramified coverings, Global analysis--papers in honor of K. Kodaira, Univ. of Tokyo Press, Princeton Univ. Press, 1969, 253-265

16. F. Hirzebruch: Pontrjagin classes of rational homology manifolds and the signature of some affine hypersurfaces, Proceedings of Liverpool Singularities Symposium II (ed. C. T. C. Wall), Lecture notes in math. 209, Springer-Verlag, Berlin-Heidelberg-New York, 1971, 207-212

17. F. Hirzebruch: The signature theorem: reminiscences and recreation, Prospects in mathematics, Ann. of Math. Studies 70, Princeton University Press 1971, 3-31

18. F. Hirzebruch: Free involutions on manifolds and some elementary number theory, Symposia Mathematica (Instituto Nazionale de Alta Matematica, Roma) Vol. V, Academic Press, 1971, 411-419

19. F. Hirzebruch and K. Jänich: Involutions and singularities, Proc. of the Bombay Colloquium on Algebraic Geometry (1968) 219-240

20. F. Hirzebruch and K. H. Mayer: O(n)-Mannigfaltigkeiten, exotische Sphären und Singularitäten, Lecture notes in math. 57, Springer-Verlag, Berlin-Heidelberg-New York, 1968

21. F. Hirzebruch and D. Zagier: Applications of the Atiyah-Bott-Singer theorem in topology and number theory, lecture notes, to be published

22. H. Lang: Über eine Gattung elementar-arithmetischer Klasseninvarianten reell-quadratischer Zahlkörper, J. für die reine und angew. Math. 233 (1968) 123-175

23. J. Lehner: Discontinuous groups and automorphic functions, Mathematical surveys VIII, American Math. Soc., Providence, 1964

24. I. G. Macdonald: The Poincaré polynomial of a symmetric product, Proc. Camb. Phil. Soc. 58 (1962) 563-568

25. I. G. Macdonald: Symmetric products of an algebraic curve, Topology 1 (1962) 319-343

26. N. Martin and C. R. F. Maunder: Homology cobordism bundles, Topology 10 (1971) 93-110

27. C. R. F. Maunder: On the Pontrjagin classes of homology manifolds, Topology 10 (1971) 111-118

28. C. Meyer: Über einige Anwendungen Dedekindscher Summen, J. für die reine und angew. Math. 198 (1957) 143-203

29. C. Meyer: Über die Bildung von Klasseninvarianten binärer quadratischer Formen mittels Dedekindscher Summen, Abhand. Math. Sem. Hamburg 27 (1964) 206-230

30. C. Meyer: Über die Dedekindsche Transformationsformel für $\log \eta(\tau)$, Abhand. Math. Sem. Hamburg 30 (1967) 129-164

31. J. Milnor: Lectures on characteristic classes, notes by J. Stasheff, Princeton 1957 (mimeographed)

32. H. Osborn: Function algebras and the de Rham theorem in PL, Bull. A. M. S. 77 (1971) 386-392

33. E. Ossa: Äquivariante Cobordismustheorie, Bonn Diplomarbeit, 1967

34. E. Ossa: Cobordismustheorie von fixpunktfreien und Semifreien S^1-Mannigfaltigkeiten, Bonn Dissertation, 1969

35. F. Pham: Formules de Picard-Lefschetz généralisées et ramifications des intégrales, Bull. Soc. Math. France 93 (1965) 333-367

36. H. Rademacher: Lectures on analytic number theory, Tata Institute of Fundamental Research, Bombay, 1954-55

37. J.-P. Serre: Groupes d'homotopie et classes de groupes abéliens, Ann. of Math. 58 (1953) 258-294

38. E. Spanier: Algebraic topology, McGraw-Hill Book Co., New York, 1966

39. R. Thom: Les classes charactéristiques de Pontrjagin des variétés triangulées, Symp. Intern. Top. Alg., 1956, 54-67, Univ. de Mexico 1958

40. C. T. C. Wall: Surgery on compact manifolds, Academic Press, London-New York, 1970

41. D. Zagier: Higher-dimensional Dedekind sums, Math. Ann., 1972 (To appear)

42. D. Zagier: The Pontrjagin class of an orbit space, Topology, 1972 (To appear)

Please turn over

Vol. 178: Th. Bröcker und T. tom Dieck, Kobordismentheorie. XVI, 191 Seiten. 1970. DM 18,–

Vol. 179: Seminaire Bourbaki – vol. 1968/69. Exposés 347-363. IV. 295 pages. 1971. DM 22,–

Vol. 180: Séminaire Bourbaki – vol. 1969/70. Exposés 364-381. IV, 310 pages. 1971. DM 22,–

Vol. 181: F. DeMeyer and E. Ingraham, Separable Algebras over Commutative Rings. V, 157 pages. 1971. DM 16.

Vol. 182: L. D. Baumert. Cyclic Difference Sets. VI, 166 pages. 1971. DM 16,–

Vol. 183: Analytic Theory of Differential Equations. Edited by P. F. Hsieh and A. W. J. Stoddart. VI, 225 pages. 1971. DM 20,–

Vol. 184: Symposium on Several Complex Variables, Park City, Utah, 1970. Edited by R. M. Brooks. V, 234 pages. 1971. DM 20,–

Vol. 185: Several Complex Variables II, Maryland 1970. Edited by J. Horváth. III, 287 pages. 1971. DM 24,–

Vol. 186: Recent Trends in Graph Theory. Edited by M. Capobianco/ J. B. Frechen/M. Krolik. VI, 219 pages. 1971. DM 18.–

Vol. 187: H. S. Shapiro, Topics in Approximation Theory. VIII, 275 pages. 1971. DM 22,–

Vol. 188: Symposium on Semantics of Algorithmic Languages. Edited by E. Engeler. VI, 372 pages. 1971. DM 26,–

Vol. 189: A. Weil, Dirichlet Series and Automorphic Forms. V, 164 pages. 1971. DM 16,–

Vol. 190: Martingales. A Report on a Meeting at Oberwolfach, May 17-23, 1970. Edited by H. Dinges. V, 75 pages. 1971. DM 16,–

Vol. 191: Séminaire de Probabilités V. Edited by P. A. Meyer. IV, 372 pages. 1971. DM 26,–

Vol. 192: Proceedings of Liverpool Singularities – Symposium I. Edited by C. T. C. Wall. V, 319 pages. 1971. DM 24,–

Vol. 193: Symposium on the Theory of Numerical Analysis. Edited by J. Ll. Morris. VI, 152 pages. 1971. DM 16,–

Vol. 194: M. Berger, P. Gauduchon et E. Mazet. Le Spectre d'une Variété Riemannienne. VII, 251 pages. 1971. DM 22,–

Vol. 195: Reports of the Midwest Category Seminar V. Edited by J.W. Gray and S. Mac Lane. III, 255 pages. 1971. DM 22.–

Vol. 196: H-spaces – Neuchâtel (Suisse)- Août 1970. Edited by F. Sigrist, V, 156 pages. 1971. DM 16,–

Vol. 197: Manifolds – Amsterdam 1970. Edited by N. H. Kuiper. V, 231 pages. 1971. DM 20,–

Vol. 198: M. Hervé, Analytic and Plurisubharmonic Functions in Finite and Infinite Dimensional Spaces. VI, 90 pages. 1971. DM 16.–

Vol. 199: Ch. J. Mozzochi, On the Pointwise Convergence of Fourier Series. VII, 87 pages. 1971. DM 16,–

Vol. 200: U. Neri, Singular Integrals. VII, 272 pages. 1971. DM 22,–

Vol. 201: J. H. van Lint, Coding Theory. VII, 136 pages. 1971. DM 16,–

Vol. 202: J. Benedetto, Harmonic Analysis on Totally Disconnected Sets. VIII, 261 pages. 1971. DM 22,–

Vol. 203: D. Knutson, Algebraic Spaces. VI, 261 pages. 1971. DM 22,–

Vol. 204: A. Zygmund, Intégrales Singulières. IV, 53 pages. 1971. DM 16,–

Vol. 205: Séminaire Pierre Lelong (Analyse) Année 1970. VI, 243 pages. 1971. DM 20,–

Vol. 206: Symposium on Differential Equations and Dynamical Systems. Edited by D. Chillingworth. XI, 173 pages. 1971. DM 16,–

Vol. 207: L. Bernstein, The Jacobi-Perron Algorithm – Its Theory and Application. IV, 161 pages. 1971. DM 16,–

Vol. 208: A. Grothendieck and J. P. Murre, The Tame Fundamental Group of a Formal Neighbourhood of a Divisor with Normal Crossings on a Scheme. VIII, 133 pages. 1971. DM 16,–

Vol. 209: Proceedings of Liverpool Singularities Symposium II. Edited by C. T. C. Wall. V, 280 pages. 1971. DM 22,–

Vol. 210: M. Eichler, Projective Varieties and Modular Forms. III, 118 pages. 1971. DM 16,–

Vol. 211: Théorie des Matroïdes. Edité par C. P. Bruter. III, 108 pages. 1971. DM 16,–

Vol. 212: B. Scarpellini, Proof Theory and Intuitionistic Systems. VII, 291 pages. 1971. DM 24,–

Vol. 213: H. Hogbe-Nlend, Théorie des Bornologies et Applications. V, 168 pages. 1971. DM 18,–

Vol. 214: M. Smorodinsky, Ergodic Theory, Entropy. V, 64 pages. 1971. DM 16,–

Vol. 215: P. Antonelli, D. Burghelea and P. J. Kahn, The Concordance-Homotopy Groups of Geometric Automorphism Groups. X, 140 pages. 1971. DM 16,–

Vol. 216: H. Maaß, Siegel's Modular Forms and Dirichlet Series. VII, 328 pages. 1971. DM 20,–

Vol. 217: T. J. Jech, Lectures in Set Theory with Particular Emphasis on the Method of Forcing. V, 137 pages. 1971. DM 16,–

Vol. 218: C. P. Schnorr, Zufälligkeit und Wahrscheinlichkeit. IV, 212 Seiten 1971. DM 20,–

Vol. 219: N. L. Alling and N. Greenleaf, Foundations of the Theory of Klein Surfaces. IX, 117 pages. 1971. DM 16,–

Vol. 220: W. A. Coppel, Disconjugacy. V, 148 pages. 1971. DM 16,–

Vol. 221: P. Gabriel und F. Ulmer, Lokal präsentierbare Kategorien. V, 200 Seiten. 1971. DM 18,–

Vol. 222: C. Meghea, Compactification des Espaces Harmoniques. III, 108 pages. 1971. DM 16,–

Vol. 223: U. Felgner, Models of ZF-Set Theory. VI, 173 pages. 1971. DM 16,–

Vol. 224: Revètements Etales et Groupe Fondamental. (SGA 1). Dirigé par A. Grothendieck XXII, 447 pages. 1971. DM 30,–

Vol. 225: Théorie des Intersections et Théorème de Riemann-Roch. (SGA 6). Dirigé par P. Berthelot, A. Grothendieck et L. Illusie. XII, 700 pages. 1971. DM 40,–

Vol. 226: Seminar on Potential Theory, II. Edited by H. Bauer. IV, 170 pages. 1971. DM 18,–

Vol. 227: H. L. Montgomery, Topics in Multiplicative Number Theory. IX, 178 pages. 1971. DM 18,–

Vol. 228: Conference on Applications of Numerical Analysis. Edited by J. Ll. Morris. X, 358 pages. 1971. DM 26,–

Vol. 229: J. Väisälä, Lectures on n-Dimensional Quasiconformal Mappings. XIV, 144 pages. 1971. DM 16,–

Vol. 230: L. Waelbroeck, Topological Vector Spaces and Algebras. VII, 158 pages. 1971. DM 16,–

Vol. 231: H. Reiter, L1-Algebras and Segal Algebras. XI, 113 pages. 1971. DM 16,–

Vol. 232: T. H. Ganelius, Tauberian Remainder Theorems. VI, 75 pages. 1971. DM 16,–

Vol. 233: C. P. Tsokos and W. J. Padgett. Random Integral Equations with Applications to Stochastic Systems. VII, 174 pages. 1971. DM 18,–

Vol. 234: A. Andreotti and W. Stoll. Analytic and Algebraic Dependence of Meromorphic Functions. III, 390 pages. 1971. DM 26,–

Vol. 235: Global Differentiable Dynamics. Edited by O. Hájek, A. J. Lohwater, and R. McCann. X, 140 pages. 1971. DM 16,–

Vol. 236: M. Barr, P. A. Grillet, and D. H. van Osdol. Exact Categories and Categories of Sheaves. VII, 239 pages. 1971. DM 20,–

Vol. 237: B. Stenström. Rings and Modules of Quotients. VII, 136 pages. 1971. DM 16,–

Vol. 238: Der kanonische Modul eines Cohen-Macaulay-Rings. Herausgegeben von Jürgen Herzog und Ernst Kunz. VI, 103 Seiten. 1971. DM 16,–

Vol. 239: L. Illusie, Complexe Cotangent et Déformations I. XV, 355 pages. 1971. DM 26,–

Vol. 240: A. Kerber, Representations of Permutation Groups I. VII, 192 pages. 1971. DM 18,–

Vol. 241: S. Kaneyuki, Homogeneous Bounded Domains and Siegel Domains. V, 89 pages. 1971. DM 16,–

Vol. 242: R. R. Coifman et G. Weiss, Analyse Harmonique Non-Commutative sur Certains Espaces. V, 160 pages. 1971. DM 16,–

Vol. 243: Japan-United States Seminar on Ordinary Differential and Functional Equations. Edited by M. Urabe. VIII, 332 pages. 1971. DM 26,–

Vol. 244: Séminaire Bourbaki – vol. 1970/71. Exposés 382–399. IV, 356 pages. 1971. DM 26,–

Vol. 245: D. E. Cohen, Groups of Cohomological Dimension One. V, 99 pages. 1972. DM 16,–